ISSUES IN ENVIRONMENTAL SCIENCE
AND TECHNOLOGY

EDITORS: R. E. HESTER AND R. M. HARRISON

5
Agricultural Chemicals and the Environment

THE ROYAL
SOCIETY OF
CHEMISTRY
Information
Services

ISBN 0-85404-220-2
ISSN 1350-7583

A catalogue record for this book is available from the British Library

Published by The Royal Society of Chemistry, Thomas Graham House,
Science Park, Milton Road, Cambridge CB4 4WF, UK

Typeset in Great Britain by Vision Typesetting, Manchester
Printed and bound in Great Britain by Bath Press, Bath

Preface

Enormous increases in agricultural productivity can properly be associated with the use of chemicals. This statement applies equally to crop production through the use of fertilizers, herbicides, and pesticides, as to livestock production and the associated use of drugs, steroids, and other growth accelerators. There is, however, a dark side to this picture and it is important to balance the benefits which flow from the use of agricultural chemicals against their environmental impacts, which sometimes are seriously disadvantageous. Within this volume we explore a variety of issues which currently are subject to wide-ranging debate and are of concern not only to the scientific establishment and to students but also to farmers, landowners, managers, legislators, and to the public at large.

In the first article, T. M. Addiscott of the Rothamsted Experimental Station examines the use of nitrogen fertilizers to increase growth rates of agricultural crops and the problems associated with nitrate leaching from soils. He reviews the potential health hazards constituted by high nitrate levels, details a number of case studies such as winter wheat production, and discusses both the science and the politics of the nitrate problem. The role of agricultural fertilizers in promoting excessive growth of aquatic weeds in rivers and of algal blooms in lakes is then discussed by A. J. D. Ferguson and M. J. Pearson of the National Rivers Authority Centre for Toxic and Persistent Substances, together with C. S. Reynolds of the Freshwater Biological Association. Their article describes problems experienced with algal blooms caused by eutrophication and examines the sources of nutrients, demonstrating that phosphate levels commonly impose a controlling influence.

The impact of agricultural pesticides on water quality is next reviewed by K. R. Eke, A. D. Barnden, and D. J. Tester, also of the National Rivers Authority. Their article outlines the legislation governing the use of pesticides in the UK and provides information on the most used chemicals, their functions, toxicities, and routes for entering water courses. The article covers recent pesticide developments and considers how pesticides will be managed in the future so as to protect the aquatic environment. D. Fowler, M. A. Sutton, U. Skiba, and K. J. Hargreaves of the Institute for Terrestrial Ecology, Penicuik, show in their article on agricultural nitrogen and emissions to the atmosphere how airborne pollutants can result from agricultural practices. They examine the conditions which lead to emissions of nitrogen oxides, mainly NO and N_2O, from soils in which vegetation and an active microbial community are present, as well as the scale of emissions of reduced nitrogen as NH_3 from animal waste. These emissions make a major contribution to the pollutant burden of the atmosphere.

Drugs and dietary additives, their use in animal production, and potential environmental consequences are discussed by T. Acamovic of the Scottish Agricultural College and C. S. Stewart of the Rowett Research Institute, both in Aberdeen. Intensive systems of livestock production for food involve major inputs of chemicals which help to preserve the health and welfare of the animals by preventing nutritional deficiencies and inhibiting disease. Some dietary additives, such as growth promoting steroids and probiotics, are designed to enhance muscle mass and to produce leaner tissue as increasingly is demanded by consumers. The residues and metabolites of such additives may be innocuous, harmful or beneficial to man and the environment and these issues are examined in depth here.

The final article, by S. G. Bell and G. A. Codd of the University of Dundee Department of Biological Services, is concerned with detection, analysis, and risk assessment of cyanobacterial toxins. These can be responsible for animal, fish, and bird deaths and for ill-health in humans. The occurrence of toxic cyanobacterial blooms and scums on nutrient-rich waters is a world-wide phenomenon and cases are cited from Australia, the USA, and China, as well as throughout Europe. The causes, indentification and assessment of risk, and establishment of criteria for controlling risk are discussed.

We believe that this collection of reviews provides an excellent treatment of some of the most important issues connected with the use of agricultural chemicals and their environmental impacts. The authors all are recognized as experts in their subjects and together they provide an authoritative and critical overview of this highly topical area of concern. The issue as a whole will be an extremely valuable source of information for all students, scientists, agricultural practitioners, consultants, managers and legislators with interest in agriculture and the environment.

Ronald E. Hester
Roy M. Harrison

Contents

Contents

Editors

Ronald E. Hester, BSc, DSc(London), PhD(Cornell), FRSC, CChem

Ronald E. Hester is Professor of Chemistry in the University of York. He was for short periods a research fellow in Cambridge and an assistant professor at Cornell before being appointed to a lectureship in chemistry in York in 1965. He has been a full professor in York since 1983. His more than 250 publications are mainly in the area of vibrational spectroscopy, latterly focusing on time-resolved studies of photoreaction intermediates and on biomolecular systems in solution. He is active in environmental chemistry and is a founder member and former chairman of the Environment Group of The Royal Society of Chemistry and editor of 'Industry and the Environment in Perspective' (RSC, 1983) and 'Understanding Our Environment' (RSC, 1986). As a member of the Council of the UK Science and Engineering Research Council and several of its sub-committees, panels, and boards, he has been heavily involved in national science policy and administration. He was, from 1991–93, a member of the UK Department of the Environment Advisory Committee on Hazardous Substances and is currently a member of the Publications and Information Board of The Royal Society of Chemistry.

Roy M. Harrison, BSc, PhD, DSc (Birmingham), FRSC, CChem, FRMetS, FRSH

Roy M. Harrison is Queen Elizabeth II Birmingham Centenary Professor of Environmental Health in the University of Birmingham. He was previously Lecturer in Environmental Sciences at the University of Lancaster and Reader and Director of the Institute of Aerosol Science at the University of Essex. His more than 200 publications are mainly in the field of environmental chemistry, although his current work includes studies of human health impacts of atmospheric pollutants as well as research into the chemistry of pollution phenomena. He is a former member and past Chairman of the Environment Group of The Royal Society of Chemistry for whom he has edited 'Pollution: Causes, Effects and Control' (RSC, 1983; Second Edition, 1990) and 'Understanding our Environment: An Introduction to Environmental Chemistry and Pollution' (RSC, Second Edition, 1992). He has a close interest in scientific and policy aspects of air pollution, currently being Chairman of the Department of Environment Quality of Urban Air Review Group as well as a member of the DoE Expert Panel on Air Quality Standards and Photochemical Oxidants Review Group and the Department of Health Committee on the Medical Effects of Air Pollutants.

Contributors

T. Acamovic, *The Scottish Agricultural College, 581 King Street, Aberdeen AB9 1UD, UK*

T.M. Addiscott, *Rothamsted Experimental Station, Harpenden, Hertfordshire AL5 2JQ, UK*

A.D. Barnden, *National Rivers Authority, National Centre for Toxic and Persistent Substances, Kingfisher House, Goldhay Way, Orton Goldhay, Peterborough PE2 5ZR, UK*

S.G. Bell, *Department of Biological Sciences, University of Dundee, Dundee DD1 4HN, UK*

G.A. Codd, *Department of Biological Sciences, University of Dundee, Dundee DD1 4HN, UK*

K.R. Eke, *National Rivers Authority, National Centre for Toxic and Persistent Substances, Kingfisher House, Goldhay Way, Orton Goldhay, Peterborough PE2 5ZR, UK*

A.J.D. Ferguson, *National Rivers Authority, National Centre for Toxic and Persistent Substances, Kingfisher House, Goldhay Way, Orton Goldhay, Peterborough PE2 5ZR, UK*

D. Fowler, *Institute of Terrestrial Ecology, Edinburgh Research Station, Bush Estate, Penicuik, Midlothian EH26 0QB, UK*

K.J. Hargreaves, *Institute of Terrestrial Ecology, Edinburgh Research Station, Bush Estate, Penicuik, Midlothian EH26 0QB, UK*

M.J. Pearson, *National Rivers Authority, National Centre for Toxic and Persistent Substances, Kingfisher House, Goldhay Way, Orton Goldhay, Peterborough PE2 5ZR, UK*

C.S. Reynolds, *Institute of Freshwater Ecology, Freshwater Biological Association, Ferry House, Far Sawry, Ambleside, Cumbria LA22 0DP, UK*

U. Skiba, *Institute of Terrestrial Ecology, Edinburgh Research Station, Bush Estate, Penicuik, Midlothian EH26 0QB, UK*

Contributors

C.S. Stewart, *Rowett Research Institute, Greenburn Road, Bucksburn, Aberdeen AB2 9SB, UK*

M. A. Sutton, *Institute of Terrestrial Ecology, Edinburgh Research Station, Bush Estate, Penicuik, Midlothian EH26 0QB, UK*

D. J. Tester, *National Rivers Authority, National Centre for Toxic and Persistent Substances, Kingfisher House, Goldhay Way, Orton Goldhay, Peterborough PE2 5ZR, UK*

Fertilizers and Nitrate Leaching

THOMAS M. ADDISCOTT

1 The Nitrate Problem

Nitrate is one of the facts of life. It is essential for the growth of many plant species, including most of those we eat, but it becomes a problem if it gets into water in which it is not wanted. It is perceived mainly as a chemical fertilizer used by farmers, but much of the nitrate found in soil is produced by the microbes that break down plant residues and other nitrogen-containing residues in the soil. There is no difference between nitrate from fertilizer and that produced by microbes, but, whatever its origin, this rather commonplace chemical entity has now become a major environmental problem and is also treated as a health hazard.

Concentrations of nitrate increased in natural waters for two decades before levelling off in the 1980s.[1] Applications of nitrogen fertilizer followed a similar pattern,[1] so many people drew the obvious conclusion that the increase in nitrate concentrations arose from the greater use of fertilizers. However, conclusions drawn from coincident changes need to be examined carefully. The fact that the birth-rate in Europe declined at the same time as the population of storks does not necessarily mean that storks bring babies! The stork and the birth-rate were probably both responding to the increased size and affluence of the human population and its increased and more intensive use of land for agriculture. Much the same can be said for nitrate concentrations and fertilizer use. It is not so much that more use of fertilizer has led to more nitrate in natural waters, as that increases in both the area of land used for arable agriculture and the intensity with which it is farmed have led to both greater concentrations of nitrate and greater use of nitrogen fertilizer. The fertilizer is part of the intensification package, and examining it as a cause of the nitrate problem without considering the rest of the package could lead to false conclusions.[2] This point is particularly important because, as we shall see later, nitrogen fertilizers have both a direct and an indirect role in the nitrate problem.

Nitrate is seen as a threat to both public health and natural waters. Of these threats the latter is definitely the more immediate, but the health issue has attracted more public concern.

---

[1] House of Lords Select Committee on the European Communities, *Nitrate in Water*, HMSO, London, 1989.

[2] T. M. Addiscott, A. P. Whitmore and D. S. Powlson, *Farming, Fertilizers and the Nitrate Problem*, CAB, Wallingford, 1991.

Nitrate as a Health Hazard

Nitrate is not a new problem. Excessive concentrations were recorded in many domestic wells in a survey conducted 100 years ago.[3] What is new is the public concern about nitrate. This arises from two medical conditions that have been linked to nitrate: methaemoglobinaemia ('blue-baby syndrome') in infants, and stomach cancer in adults. Both are serious conditions, so we need to examine possible links carefully, but we need to note that these conditions are not caused by nitrate but by the nitrite to which it may be reduced. Nitrate itself is harmless and is most notable from a medical standpoint as a treatment for phosphatic kidney stones.

Methaemoglobinaemia. The 'blue-baby syndrome' can occur when an infant less than about one year old ingests too much nitrate. Microbes in the stomach convert the nitrate to nitrite and when this reaches the blood-stream it reacts with the haemoglobin, the molecule that transports oxygen around the body. Normal oxyhaemoglobin, which contains iron in the iron(II) state, becomes methaemoglobin in which the iron is in the iron(III) state, greatly lessening the capacity of the blood to carry oxygen and causing what might be described as chemical suffocation. Very young children are susceptible because foetal haemoglobin, which has a greater affinity for nitrite than normal haemoglobin, persists in the blood-stream for a while, and because their stomachs are not sufficiently acid to inhibit the microbes that convert nitrate to nitrite. Gastroenteritis greatly exacerbates the effects of the nitrite.

This condition is usually very rare. In the UK the last case was in 1972 and the last death[4] in 1950, but in Hungary there were over 1300 cases between 1976 and 1982.[5] One reason for this difference between the two countries may lie in the origin of the water. Practically all known cases were associated with water from wells, and the condition is known as 'well-water methaemoglobinaemia' in the USA. In 98% of these cases the wells were dug privately and may have been too close to disposal points for animal or human excreta, thereby increasing the risk of pollution not only by nitrate but also by *E. coli* and other organisms that cause gastroenteritis. The author is not aware of any case in which methaemoglobinaemia was caused by tap water from a mains supply such as that used by most households in the UK. That being said, there is no room for complacency about the condition. In the fatal case in 1950, the doctor reported that, 'There were diarrhoea and vomiting and the child's complexion was slate-blue'.[4] In a similar but non-fatal case in the same year, 'Blood drawn from a vein was a deep chocolate-brown'.[4]

Stomach Cancer. Of all the cancers, that of the stomach causes the second largest number of deaths. Only lung cancer kills more men, and only breast cancer kills more women. Stomach cancer is a painful and debilitating way to die, and the link to nitrate in water that has been suggested is a serious matter. There

---

[3] R. Warington, *J. Chem. Soc.*, 1887, 500.

[4] M. C. Ewing and R. M. Mayon-White, *Lancet*, 1951, **260**, 931.

[5] S. Deak, quoted in *Health Hazards from Nitrates in Drinking Water*, Report on a WHO meeting, Copenhagen, 5–9 March 1984, WHO, Geneva, 1985.

are good theoretical reasons for proposing such a link. Nitrite produced from nitrate could react in the stomach with a secondary amine coming from the breakdown of meat or other protein to produce an N-nitroso compound.

$$\frac{R^1}{R^2}N-H + NO_2^- + H^+ \longrightarrow \frac{R^1}{R^2}N-N=O + H_2O \tag{1}$$

The N-nitroso compounds are carcinogenic, so the reaction could result in stomach cancer. This mechanism is essentially a hypothetical one, and tests were made to evaluate it. Three tests in particular suggested that the hypothesis was not correct, that is, that there is no clear link between stomach cancer and nitrate in water.

One test, made at the Radcliffe Infirmary in Oxford,[6] identified two areas of the UK in which the incidence of stomach cancer was particularly high and two in which it was particularly low. People attending hospitals in these areas as visitors rather than patients were asked to provide samples of saliva. The hypothesis suggested that samples from the high-risk areas should contain more nitrite and nitrate than those from the low-risk areas, but this was not so. The samples from the low-risk populations had nitrate concentrations 50% greater than those from the high-risk areas.

Another test looked at nitrate concentrations in 229 urban areas in the UK between 1969 and 1973 and at deaths from stomach cancer in the same areas at the same time.[7] The hypothesis suggested that there should be a positive relationship between them. The results showed a negative one.

The third test is particularly interesting in the context of a paper about nitrogen fertilizer because it involved a plant making it.[8] If you work in, or even visit, a plant producing ammonium nitrate or another fertilizer, you can taste fertilizer within minutes of entering, as the fine dust in the air dissolves in your saliva. If anyone is going to get stomach cancer as a result of exposure to nitrate, it is workers in nitrate fertilizer plants, but an epidemiological survey showed that their mortality due to stomach cancer did not differ from that among workers in comparable jobs (Table 1). Twelve deaths occurred from stomach cancer during the 35 years of the study, while 12.06 deaths would have been expected from the mortality rate among the other workers.

Although there are a few caveats that need to be noted about these tests,[2] the results suggest strongly that there is no real link between stomach cancer and nitrate in water. Further support for this conclusion comes from the fact that, while nitrate concentrations have been increasing in water during the past 30 years, the incidence of stomach cancer has been declining, and about ten years ago the absence of any link was accepted officially.[9]

Is there a Safe Limit for Nitrate?   The two previous sections show that any safe limit for nitrate needs to be related to methaemoglobinaemia rather than

[6] D. Forman, S. Al-Dabbagh and R. Doll, *Nature (London)*, 1985, **313**, 620.
[7] S. A. Beresford, *Int. J. Epidemiol.*, 1985, **14**, 57.
[8] S. Al-Dabbagh, D. Forman, D. Bryson, I. Stratton and R. Doll, *Br. J. Ind. Med.*, 1986, **43**, 507.
[9] E. D. Acheson, *Nitrate in Drinking Water*, HMSO, London, CMO(85), 14, 1985.

Table 1 Mortality from cancer of the stomach and other causes between 1 January 1946 and 28 February 1981 among 1327 male workers in a factory making ammonium nitrate fertilizer. Deaths observed compared with deaths expected from statistics for workers in comparable jobs in the locality. Statistics for heavily and less heavily exposed workers combined*

Disease[†]	Number of deaths	
	Observed	Expected
Stomach cancer	12	12.06
All cancers	91	86.83
Respiratory diseases[‡]	36	51.04
Heart disease[‡]	92	113.72
All causes	304	368.11

*Taken from Al-Dabbagh *et al.*[8]

[†]The differences for cancer were not statistically significant, but those for the other two categories of disease were ($p < 0.05$; *i.e.* only a 1 in 20 chance that the difference was accidental). For all causes, $p < 0.001$; *i.e.* only a 1 in 1000 chance that the difference was accidental.

[‡]The large difference in the mortality from respiratory and heart disease may have a simple explanation. Ammonium nitrate is explosive and the fertilizer workers were probably not allowed to smoke at work.

stomach cancer. 'Blue-baby' cases in the USA were associated[2] with nitrate concentrations ranging from 283 to 1200 g m^{-3}. The only fatal case in the UK arose from a nitrate concentration of 200 g m^{-3}. The non-fatal case in the same report involved bacterially polluted water with 95 g m^{-3} of nitrate. There is no obvious safe level, but in evidence to the House of Lords[1] three august bodies, The Medical Research Council, The Institute of Biology, and the Institute of Cancer Research, stated that the majority of cases have occurred when the water contained more than 100 g m^{-3} of nitrate. The limit of 50 g m^{-3} imposed by the European Commission thus has a safety factor of two. This safety factor is a mixed blessing. To some it represents prudence, but to farmers and water suppliers it is a problem. A limit of 100 g m^{-3} for water draining from agricultural land would be quite easy to achieve, but 50 g m^{-3} is far more difficult,[1] particularly in the drier areas of the UK and when there is a substantial input of nitrogen to the soil from the atmosphere.

Nitrate as an Environmental Problem

It is not only land plants that need nitrogen for growth. Plants growing in water respond to extra nitrogen like crop plants, but their extra growth is not welcome. Increased nitrate concentrations in rivers and lakes may encourage reeds to grow to excess, narrowing waterways and possibly overloading and damaging banks.[10] Underwater plants also proliferate, so that anglers lose tackle, the propellers of boats get fouled and conduits for water supply become clogged and machinery damaged.

Large water plants, though a nuisance, are not the main environmental problem. Algae are very small single-celled plants that grow on a variety of surfaces, including that of water. They are not noticeable on water until they grow to excess and form the 'blooms'—possibly better described as scum—that are

[10] The Royal Society, *The Nitrogen Cycle of the United Kingdom: A Study Group Report*, The Royal Society, London, 1983.

now seen to a worrying extent on our rivers and lakes. These blooms look messy when they grow, but they become a far bigger problem when they die. The bacteria that decompose them use oxygen when they do so and thus deprive fish and other desirable organisms, which may die. The whole ecological balance of the river or lake may change, usually to the detriment of species that we would like to see there. This process, known as 'eutrophication', is limited by phosphate rather than nitrate in fresh water, and it is discussed in detail elsewhere in this issue.[11] Algae can also grow to excess in the sea, but these marine blooms seem to be stimulated more by nitrate than by phosphate. Algal blooms took on a new significance a few years ago when it was discovered that some species were toxic to humans and dogs, a problem that is again discussed in more detail elsewhere in this issue.[12] The indirect effects of nitrate in water proved in this instance to be a far more tangible health hazard than the direct effects.

2   The Contribution of Fertilizer to the Nitrate Problem

At the root of the nitrate problem lies a very simple relationship[2]

$$availability = vulnerability$$

Any nitrogen in the soil that is available to crops is likely to be present as nitrate itself, or as ammonium, which microbes in the soil soon convert to nitrate. Nitrate is completely soluble in water in the presence of all cations likely to be in the soil solution and it is not adsorbed. It is thus vulnerable to being washed out of the soil by percolating rainfall or irrigation. The surest way of avoiding such nitrate losses is to ensure that as little nitrate as possible is in the soil at any time. When crops are growing fast they take up nitrogen quickly, as much as $5\,kg\,ha^{-1}\,d^{-1}$ sometimes, so they need a generous supply of nitrate in the soil. Once they cease to grow and to take up nitrate, however, we need to make sure that there is as little nitrate as possible left in the soil so that it is not vulnerable to leaching. Any nitrate present is there at the wrong time and we can see that what we have is not so much a nitrate problem as a problem of untimely nitrate.[2] Most untimely nitrate happens because either the soil microbes or the farmer have put it in the soil when the crop cannot use it. However, there is another scenario that occurs when the farmer applies the fertilizer at a perfectly sensible time, but substantial amounts of rain fall before the crop has had a chance to use the nitrogen supplied. To assess the contribution of nitrogen fertilizer to the nitrate problem, we need to think about how it might become untimely nitrate. In the first few months after application the four most likely fates for the fertilizer are:

1. It may be taken up by the crop, as intended.
2. It may become incorporated in the soil's organic matter, where it will remain unless it is remobilized by the bacteria and other organisms in the soil.
3. It may be leached out of the soil.
4. It may be denitrified. This happens when microbes hungry for oxygen utilize

[11] A. J. D. Ferguson, M. J. Pearson and C. S. Reynolds, this *Issue*, p. 27.
[12] S. G. Bell and G. A. Codd, this *Issue*, p. 109.

the oxygen atoms of the nitrate ion so that NO_3^- becomes dinitrogen (N_2) or nitrous oxide (N_2O), both of which are gases. Denitrification lessens the nitrate problem but N_2O contributes to the 'greenhouse effect'. Ammonium in the fertilizer may be volatilized as ammonia.

Fertilizer nitrogen is usually applied in the spring, but to assess its overall contribution to the nitrate problem we need to consider what happens not only in the period between its application and the harvest of the crop but also after harvest.

Birds and animals are 'tagged' so that their wanderings and ultimate fate can be followed. To do the same for nitrogen fertilizer, we use the heavy isotope of nitrogen, ^{15}N, as a 'tag' or 'label'. This is a safe isotope to use because it is not radioactive. The ^{15}N of the label is taken up by the crop, incorporated in the soil's organic matter or leached or denitrified in almost exactly the same way as the ^{14}N which makes up 99.6% of the nitrogen in the rest of the fertilizer and in the soil. The ratio of ^{15}N to ^{14}N in the fertilizer is known, so when the crop and the soil are analysed and the ^{15}N to ^{14}N ratios are determined (usually by mass spectrometry), the amounts of nitrogen of fertilizer origin that are in the crop or in the organic matter, ammonium, or nitrate in the soil can be determined.

The fact that ^{15}N is not radioactive means that it can be used safely in experiments in the field, but it also means that much patient work is needed to obtain results. The approach is demanding in terms of time, equipment, and skilled manpower, but it has made a great contribution to the understanding of the nitrate problem. The results that are outlined here are from experiments made by staff at Rothamsted,[13,14] but key contributions have also come from Scotland[15] and France.[16] The majority of the Rothamsted experiments involved winter wheat, but oilseed rape, potatoes, beans, and sugar beet were also grown.[17] The soil is a factor in nitrate leaching, and three types were used, the flinty, silty clay loam at Rothamsted, a sandy loam at Woburn in Bedfordshire and a heavy sandy clay at Saxmundham in Suffolk.

Winter Wheat—the Baseline

Winter wheat formed the backbone of the ^{15}N programme for several reasons. One is simply that it is the most widely grown crop in England and Wales. Winter cereals occupy about 60% of the arable land in England. Winter wheat is also representative of those crops which are sown in the autumn and have a well-established root system by the time they receive applications of nitrogen fertilizer in the spring. It is a notably parsimonious crop—only sugar beet allows less nitrate to escape its roots—so it provides a useful baseline against which to compare other crops.

[13] D. S. Powlson, P. B. S. Hart, P. R. Poulton, A. E. Johnston and D. S. Jenkinson, *J. Agric. Sci., Camb.*, 1992, **118**, 83.

[14] A. J. Macdonald, D. S. Powlson, P. R. Poulton and D. S. Jenkinson, *J. Sci. Food Agric.*, 1989, **46**, 407.

[15] K. A. Smith, A. E. Elmes, R. S. Howard and M. F. Franklin, *Plant Soil*, 1984, **76**, 49.

[16] S. Recous, C. Fresneau, G. Faurie and B. Mary, *Plant Soil*, 1988, **112**, 215.

[17] A. J. Macdonald, P. R. Poulton and D. S. Powlson, in *Proceedings of the First Congress of the European Society of Agronomy, Paris, 1990*. European Society for Agronomy, Colmar, France.

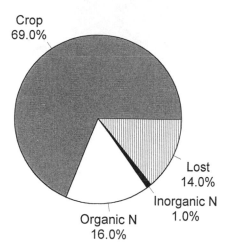

ure 1 Fate of ^{15}N-labelled
ogen fertilizer applied to
inter wheat crop. (Taken
rom Macdonald *et al.*[14])

Crop
69.0%

Lost
14.0%

Inorganic N
1.0%

Organic N
16.0%

Between Fertilizer Application and Harvest. The results from Rothamsted showed that 50–80% of the labelled fertilizer nitrogen was recovered in the crop, while a further 10–25% was in the soil when the crop was harvested.[13,14] The key finding was that almost all the labelled nitrogen in the soil was in organic forms (Figure 1). Some will have been in dead roots and some in living matter, such as microbes and small soil animals that will have fed on the roots and on exudates produced by the roots while they were still alive. The proportion of the labelled nitrogen left vulnerable to leaching as ammonium or nitrate was only 1–2%, and nearly all the nitrate found in the soil at harvest was not labelled and therefore did not come from the fertilizer.

The best overall recovery of labelled fertilizer in crop and soil was 99% and the least satisfactory 65%.[13] Thus 1–35% of the labelled nitrogen, on average 15.7%, was 'missing, presumed lost'. These losses occurred between the time of the fertilizer application in spring and the time the crop was harvested; but why and how did they occur?

The reason why the losses occurred is simply 'because it rained'. The more rain that fell in the period after application, the more labelled nitrogen was lost. The critical period seemed to be the first three weeks after application; this was the period that showed the closest relationship between the losses and the rainfall (Figure 2).[13]

How the losses occurred is also important. Rain may cause denitrification as well as leaching, but only the leaching losses cause the nitrate problem. It was not feasible at the time to measure either kind of loss directly, so we partitioned the losses between leaching and denitrification by estimating the loss by leaching with the aid of computer models and assuming that the remaining loss was by denitrification.[18] The apparent loss by denitrification, L_D, could have been an artefact caused by underestimating the loss by leaching, so it was subjected to a check. This showed that, as would be expected of denitrification, L_D was significantly better related to the computed soil wetness during the three-week

[18] T. M. Addiscott and D. S. Powlson, *J. Agric. Sci., Camb.*, 1992, **118**, 101.

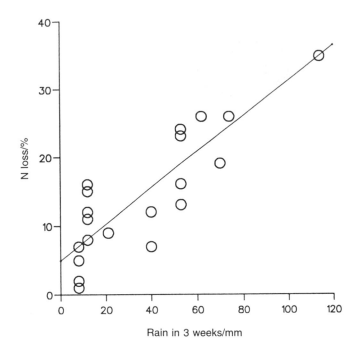

Figure 2 Losses of
[15]N-labelled nitrogen
fertilizer related to rainfall
in the three weeks after
application. (Drawn by
Addiscott *et al.*[2] from data
of Powlson *et al.*[13])

critical period than to the rainfall. An underestimate of leaching would have been
better related to the rainfall.

Losses of labelled nitrogen were partitioned in this way for 13 experiments,
suggesting that the loss was solely by leaching in two experiments and solely by
denitrification in another, but that both processes contributed in the other ten
experiments (Figure 3). Denitrification predominated overall; of the average total
loss of 16%, it contributed 10% and leaching 6%.[18] Losses by denitrification
were thus nearly twice as large as those by leaching, but it is possible that other
gaseous losses, such as volatilization of ammonia, may have contributed.

The percentage of the labelled fertilizer that was computed to have been
leached was fairly small, 7% or less, in 11 of the 13 experiments. Two much larger
losses were computed, one of 19% when 70 mm of rain fell in three weeks on the
light sandy loam, and one of 23% when 114 mm of rain fell in three weeks on the
heavy, sandy clay. We need to ask, however, how close these computed losses are
likely to be to reality. The Brimstone Farm Experiment[19] measures losses of
nitrate by leaching from plots of heavy clay soil growing winter wheat and other
winter-sown crops. The losses of nitrate-nitrogen between fertilizer application
and harvest derive from nitrate in the soil as well as from fertilizer,[20] but it is
useful to express them as percentages of the nitrogen applied. The mean
percentage loss obtained by this calculation was 3.3% and the range was 0–17%.
The computed losses from fertilizer had a mean of 5.7% and a range of 0–23%.
Thus, although we are not really comparing like with like, the computed losses do

[19] G. L. Harris, M. J. Goss, R. J. Dowdell, K. R. Howse and P. Morgan, *J. Agric. Sci., Camb.*, 1984,
       102, 561.
[20] M. J. Goss, K. R. Howse, P. W. Lane, D. G. Christian and G. L. Harris, *J. Soil Sci.*, 1993, **44**, 35.

Figure 3 Partitioning of losses of nitrogen fertilizer between denitrification and leaching in 13 experiments. (Taken from Addiscott and Powlson,[18] with permission of Cambridge University Press.)

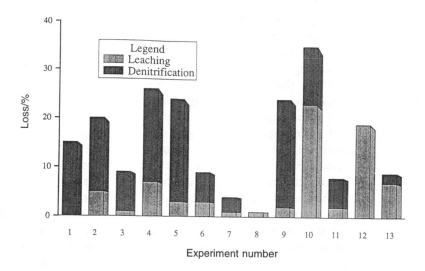

not look to be too far from reality and they seem to err on the side of pessimism.

We have seen that only 1–2% of labelled nitrogen fertilizer remains vulnerable to leaching when winter wheat is harvested and that, on average, 5–6% is leached before harvest. This implies that 6–8% of an average nitrogen application of $190 \, \text{kg ha}^{-1}$, that is $11–15 \, \text{kg ha}^{-1}$ of nitrogen, is lost by leaching. We know, however, that three to four times as much nitrate-nitrogen as this, and sometimes more, is often leaching in practice. From where does it come? To answer this question we need to ask what happens in the soil after the winter wheat has been harvested.

After Harvest. Natural systems of vegetation usually have small concentrations of nitrate in the water draining from their soils. They have evolved to fit a particular niche in the most efficient way possible, and they let little go to waste. Harvest is the point at which arable land becomes most different from a natural system of vegetation, because the soil is bare and will shortly be cultivated. No plants are there to take nitrate from the soil, so any nitrate that is there is, by definition, untimely nitrate. This is one reason for paying particular attention to the post-harvest period. Another is the changing balance between rainfall and evaporation. The soil is usually dry at harvest because of transpiration through the crop and direct evaporation during summer; however, during autumn, rainfall gradually overtakes evaporation and the soil becomes moister and eventually permits water to flow downwards, carrying nitrate with it. This is clearly a time at which we should look for the nitrate losses that we could not account for in the period before harvest.

The main agents of these losses are the microbes and small animals, such as springtails and mites, that inhabit the soil. These feed on organic matter that contains carbon and nitrogen and produce carbon dioxide and ammonium ions as waste products. Other bacteria convert the ammonium to nitrate. Like most of us, these organisms are most active when the conditions suit them best, and their preferred options are warmth and moisture. In early autumn, the soil is still warm

9

from summer, and is becoming moister, giving ideal conditions for them to produce ammonium and nitrate just when it is most untimely, in bare soil through which rain will soon be percolating for four to six months. This naturally occurring nitrate is usually responsible for a much larger proportion of the nitrate problem than the fertilizer nitrogen given in spring. In the Brimstone Farm Experiment, for example, post-harvest losses of nitrate-nitrogen were roughly five times greater than those between fertilizer application and harvest.[20]

Further support for the idea that nitrogen released from organic matter by microbes makes a major contribution to nitrate losses from the soil is given by the results of a very old experiment at Rothamsted.[21] The Drain Gauges were constructed in 1870 to measure losses of water from the soil, but nitrate was found in the drainage and was measured on a daily basis from 1877 until 1915.[22,23] (The short-term project had not yet been invented!) During the first seven years of the nitrate measurements the soil leaked an average of $45 \, kg \, ha^{-1}$ of nitrogen as nitrate each year.[22] The soil in the gauges carried no crop and it received only the cultivation needed to remove weeds. It had received no fertilizer since 1868, so those seven years were the 9th to the 16th since the last application. Only $3–5 \, kg \, ha^{-1}$ of nitrogen was brought in by rain, so almost all of the remaining $40–42 \, kg \, ha^{-1}$ must have come from the organic matter. This was confirmed by the fact that the total amount of nitrogen in the drainage, corrected for what came in rain during the whole 38 years, $1450 \, kg \, ha^{-1}$, was only a little larger than the measured decline in total nitrogen in the soil of the guages, $1410 \, kg \, ha^{-1}$. The most likely explanation for the disparity is nitrogen deposited from the atmosphere other than in rain. Mineral nitrogen in the soil is a minute fraction of the total, so practically all these losses must have come from the organic matter.

The annual loss of nitrate from the Drain Gauges declined over the 38 years but it did so very slowly (Figure 4). Removing the effects of the fluctuations in annual rainfall and fitting an exponential function[2,24] showed that it would have taken 41 years for the loss to fall to half its initial rate. It is interesting to note too that the water draining from this unfertilized soil in the 1870s did not initially conform to the EC's nitrate limit of $50 \, g \, m^{-3}$. Seven years were to pass before it would become officially potable.

The Drain Gauge results show that the microbes and other soil organisms are well able to produce nitrate without our help. We give them considerable encouragement, however, by cultivating the soil and breaking up the larger aggregates and clods so that extra organic matter is exposed to the air and thence to attack by aerobic bacteria. Tilled soil contains more nitrate in autumn than untilled soil, other things being equal.[25]

How much does Nitrogen Fertilizer Contribute to Nitrate Leaching when Winter Wheat is Grown? We saw earlier that direct nitrate losses from nitrogen fertilizer given to winter wheat are often relatively small. We saw too that

[21] J. B. Lawes, J. H. Gilbert and R. Warington, *J. R. Agric. Soc. England*, 1882, **18**, 1.
[22] N. H. J. Miller, *J. Agric. Sci., Camb.*, 1906, **1**, 377.
[23] E. J. Russell and E. H. Richards, *J. Agric. Sci., Camb.*, 1919, **10**, 22.
[24] T. M. Addiscott, *Soil Use Mgmt.*, 1988, **4**, 91.
[25] R. J. Dowdell and R. Q. Cannell, *J. Soil Sci.*, 1975, **26**, 53.

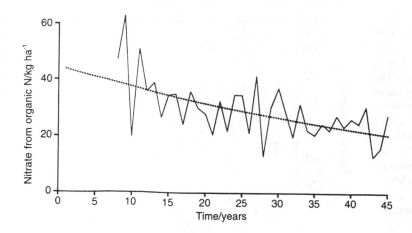

microbes and other soil organisms produce much of the nitrate that is leached from the soil in the autumn and winter. Does this mean that, when winter wheat is grown, nitrogen fertilizer is largely exonerated with respect to nitrate leaching? The large acreage given over to winter wheat makes this an important question, but there are two issues that need to be discussed before a firm answer can be given: 'memory effects', and excessive applications of nitrogen fertilizer.

The idea behind the term 'memory effects' is simple. Does the quantity of nitrogen fertilizer given to the crop in spring influence the production of nitrate by soil organisms in the autumn after the crop has been harvested? It could be, for example, that giving more nitrogen increases the amounts of easily decomposable nitrogen-containing crop residues in the soil in autumn, so that more nitrate is produced. This is an important question, because experiments at Rothamsted have shown that the 'surplus nitrate curve',[2] which relates the amount of nitrate in the soil at harvest to the amount of fertilizer given (Figure 5), often shows no increase in surplus nitrate until quite large applications of fertilizer are given. This suggests that fertilizer applications are safe until they cause an upturn in surplus nitrate. This would, however, be an erroneous conclusion if it were found that, although an application has caused no increase in surplus nitrate at harvest, it has led to increased surplus nitrate in the soil after harvest. Research at Rothamsted has shown that winter wheat grown for just one year does not show a 'memory effect' from its fertilizer application. Not surprisingly, perhaps, winter wheat grown with the same fertilizer application for 150 years does leave a definite 'memory effect'.[26]

Whether or not there is a 'memory effect' becomes rather academic if the farmer is applying excessive amounts of nitrogen fertilizer in any case. Looking at the surplus nitrate curve shows that giving more than a certain amount of nitrogen will cause a sharp upturn in the amount of nitrogen left in the soil at harvest and vulnerable to leaching. Results from the Broadbalk experiment at Rothamsted suggest that the safe limit is about 200 kg ha^{-1} of nitrogen.[13] Larger applications than this are quite common, but they are only justifiable if the crop uses them.

[26] M. J. Glendining, P. R. Poulton and D. S. Powlson, *Aspects Appl. Biol.*, 1992, **30**, 95.

Figure 5 A 'surplus nitrate curve', relating the amount of nitrate left in the soil at harvest to that given in spring as fertilizer. The point at which the curve turns upwards represents the safe limit for fertilizer application. It could be described as the environmental optimum application. Fortunately, it is often close to the economic optimum.

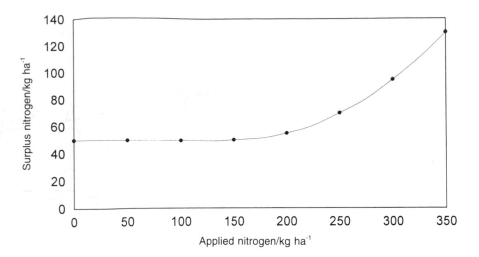

There is a useful rule-of-thumb which says that winter wheat will take up about 23 kg of nitrogen for each tonne of grain produced. Quite a few farmers achieve wheat yields of $10 \, t \, ha^{-1}$ fairly often and can make a reasonable case for applying $230 \, kg \, ha^{-1}$ of nitrogen if the contribution from soil mineral nitrogen is small, but applications of $300 \, kg \, ha^{-1}$ are not uncommon and certainly do not produce $13 \, t \, ha^{-1}$ yields every time. The crop that fails to come up to expectations is something of an environmental hazard. If the farmer aims for $10 \, t \, ha^{-1}$ and achieves only $7 \, t \, ha^{-1}$, a lot of nitrogen could be sitting unused in the soil at harvest. If the through drainage in the following winter is not very great, that unused fertilizer could on its own take the nitrate concentration in the drainage over the EC limit of $50 \, g \, m^{-3}$ of nitrate. If farmers wish to avoid draconian legislation over the nitrate problem they clearly need to think very hard before using more than $220 \, kg \, ha^{-1}$ of nitrogen.

Crops other than Winter Wheat

Although winter wheat is grown on more land than any other crop in England and Wales, it is not usually grown continuously, because of problems from pests and diseases. 'Break crops' are grown to interrupt the build-up of these problems, so we need to look at the behaviour of nitrogen under these and other crops, but much less information is available than for winter wheat.

Up to Harvest. Oilseed rape and field beans are used as break crops for winter wheat on a variety of soils, and potatoes are used on the lighter soils. Sugar beet may also be grown, but this depends not only on the soil but also on the proximity of a sugar beet processing factor. Four Rothamsted-based experiments[17] compared the effectiveness of winter wheat and winter oilseed rape in their use of labelled nitrogen fertilizer. Potatoes were included in two of these experiments and sugar beet and field beans in one experiment each. Two criteria based on the

Table 2 Efficiency of use of nitrogen fertilizer by winter wheat, winter oilseed rape, potatoes and sugar beet. Percentage of labelled fertilizer nitrogen left in soil at harvest, and percentage increase in soil mineral nitrogen at harvest caused by fertilizer*

	Winter wheat	W. oilseed rape	Potatoes	Sugar beet	Field beans
No. of experiments[†]	4	4	2	1	1
N application ($kg\,ha^{-1}$)	211	235	223	122	—
% labelled N in soil at harvest	3.5 (2.6[‡])	4.3	7.2	1.1	—
N_{min} in soil at harvest, no fertilizer ($kg\,ha^{-1}$)	58	42	40	19	86
% increase in harvest N_{min} from fertilizer	14	69	98	0	—

*Derived from Macdonald et al.[17]
[†]The data in the table were obtained by averaging across the number of experiments shown.
[‡]Omitting one crop which was seriously damaged by disease.

results can be used to assess the efficiency of fertilizer use: (1) the percentage of labelled fertilizer nitrogen left in the soil at harvest as mineral nitrogen (defined as ammonium-N + nitrate-N); (2) the percentage increase in mineral nitrogen (labelled or unlabelled) at harvest that results from the use of nitrogen fertilizer. The first measures specifically the efficiency of fertilizer use while the second indicates the efficiency with which the crop uses nitrate, regardless of its origin. Both criteria showed the same trend (Table 2). Oilseed rape seemed less efficient in its use of nitrogen fertilizer than winter wheat, while potatoes were less efficient still. Sugar beet was clearly the most efficient crop of all. Field beans do not need fertilizer nitrogen because they are legumes and can obtain nitrogen from the air through their association with nitrogen-fixing bacteria in nodules on their roots. They nevertheless left more mineral nitrogen in the soil than any other crop also not given fertilizer.

Data of a different kind, the nitrate losses through the field drains of the Brimstone experiment,[20] also suggest that winter oilseed rape uses nitrogen fertilizer somewhat less efficiently than does winter wheat, but this experiment did not include the other crops discussed above.

After Harvest. How do the 'memory effects' shown by the other crops compare with those of winter wheat? Winter wheat did not show a memory effect after one year, but oilseed rape does seem to do so. Researchers of the Agricultural Development and Advisory Service found that nitrate production by microbes in the soil after a rape crop increased with the amount of fertilizer given to the crop (R. Sylvester-Bradley, personal communication). One reason may lie in this crop's habit of shedding its leaves as harvest approaches, which means that the microbes in the soil get early access to these residues. This habit might contribute to the apparently smaller efficiency of this crop in using nitrogen fertilizer. The crop may be just as efficient as winter wheat at taking up the fertilizer but drops leaves in time for the labelled nitrogen in the leaves to be converted to mineral nitrogen by the microbes, so that it appears to be unused fertilizer nitrogen. With potatoes and sugar beet, the 'memory effects' will probably be dominated by the way in which the potato haulm and sugar beet leaves are treated. If they are ploughed into the soil in autumn, quite strong memory effects could be expected, but if they are removed, such effects are unlikely.

Application of Nitrogen Fertilizer in Autumn. Winter-sown crops were assumed for many years to need a small amount of nitrogen fertilizer in the seed-bed in autumn 'to help them through the winter'. Concern about naturally occurring nitrate in the soil in autumn brought this practice into question, and the effect of autumn-applied nitrogen on nitrate losses from the soil during autumn and winter was investigated in the Brimstone Farm experiment.[20] The results were striking. For every kilogram of nitrogen applied in autumn, an extra kilogram of nitrogen was leached as nitrate. It was probably not the fertilizer nitrogen that was leached but soil nitrogen that would have been taken up by the crop if the fertilizer had not been there. These results and others have led to a decline in the application of nitrogen in autumn in recent years. A few farmers say that they have started the practice again, and the reason illustrates the complexity of the nitrate problem. The burning of cereal straw in autumn is now banned, and farmers have either to plough it into the soil or find a use for it elsewhere. Many plough it in, but when it is ploughed into soil it immobilizes mineral nitrogen. Straw has a nitrogen to carbon ratio that is too low for the microbes to be able to use it satisfactorily, so they use mineral nitrogen from the soil to help them metabolize it.[27] This mineral nitrogen thus becomes immobilized in organic matter and is not available to crops at the beginning of autumn. The amounts involved are not huge—one tonne of straw immobilizes about 10 kg of nitrogen—but farmers are aware of this phenomenon and they are sensitive to the appearance of their crops. They like them to look green. It will be a pity if the need to plough in straw causes a resurgence in the use of nitrogen fertilizer in autumn, but the author is not aware that this is happening on a wide scale. Immobilization may well be beneficial, of course, where the farmer applied too much fertilizer nitrogen in the spring.

Another twist in the autumn nitrogen saga is worth mentioning. A few years ago, the author talked to farmers when they came to look at a display at an agricultural show. He asked them if they were still applying nitrogen in autumn. 'No', said most of them, 'except to winter oilseed rape'. Why just to oilseed rape? 'Because of the pigeons'. There is nothing pigeons like better in their autumn diet than tender young oilseed rape plants, and a large flock can do great damage. Pigeons, however, like aircraft pilots, do not like to land in heavy vegetation, and the farmers reckoned that giving autumn nitrogen helped the oilseed rape to grow quickly to a height at which it became a hazard to 'aviation' and, therefore, safe from attack. This autumn nitrogen would not otherwise have given any yield benefit.

---

[27] D. S. Jenkinson, in *Straw, Soils and Science*, ed. J. Hardcastle, AFRC, London, 1985, p. 14.

Horticultural Crops. There are at least 25 horticultural crops, so it is not possible to consider any of them individually. Most are high-value crops, so the grower can afford to use plenty of nitrogen fertilizer, and often does. The largest amounts of mineral nitrogen that the author has seen recorded in the soil after harvest have followed horticultural crops. Part of the reason for the large applications of nitrogen fertilizer made to these crops is the rather limited root systems that some of them have. To obtain good initial growth, the grower may need to apply much more nitrogen fertilizer than the crop will ultimately take up. A useful counter-measure to this problem has been the 'starter dose',[28] a relatively small application of fertilizer made close to the roots of the crop that has the same effect on growth as a much larger generalized application. Once the starter dressing has enabled the crop to establish itself, the remaining applications can be related more closely to its uptake of nitrogen.

Even when horticultural crops leave substantial amounts of nitrate in the soil at harvest, they are not usually a very important factor in the nitrate problem because horticulture occupies only about 6% of the cultivable land. However, there could be a problem if, because of the type of soil, a number of market gardens were concentrated above an aquifer that was an important source of potable water.

Crop Sequences. With both arable and horticultural crops, the nitrate problem is influenced not just by the crops grown but also by the order in which they are grown. This issue is largely one of common sense. If you grow a crop, such as potatoes, that is likely to leave a fair amount of nitrate in the soil in autumn, it pays to follow it with winter wheat or another crop that will take up that nitrate as soon as possible. This is sometimes easier in theory than in practice. It is not unknown for a farmer to experience such a wet autumn that he finds himself harvesting potatoes on Christmas Day and sowing a spring cereal later. In general, crop sequences need to be planned so that they leave the soil bare for as little time as possible. Maize is a problem in the USA and other places where it is grown, because it grows slowly and leaves the soil almost bare for a long time in spring.

Arable farmers in the UK rarely grow more than one crop in a year, but horticulturalists may grow two or more crops in a year on the same soil, so the question of crop sequences is more intense for them, particularly as *Brassica* vegetables, such as cabbages and brussel sprouts, can leave large organic as well as mineral nitrogen residues.

Grassland

The 18 million hectares of land in agricultural use in the UK in 1990 were, in rough terms, split three ways: one-third each to arable farming, managed grassland and rough grazing. About half the fertilizer used in the UK goes to the managed grassland, so this type of farming is very much part of our enquiry. Some of this managed grassland is long-term or permanent, which has obvious

[28] D. A. Stone and H. R. Rowse, *Aspects Appl. Biol.*, 1992, **30**, 399.

advantages in the prevention of nitrate problems because the soil is always covered with vegetation. Conditions that favour the production of nitrate by soil organisms also favour the uptake of nitrate by the grass and other species growing in the soil, so losses of naturally produced nitrate tend to be small. Grassland can also make efficient use of nitrogen applications of up to $400 \, kg \, ha^{-1}$ each year without any appreciable leakage of nitrate. In terms of the nitrate problem it is a model crop—until you use it. We are concerned here mainly with grazed grassland, but brief mention is made of the problems that arise from systems in which grass is cut and fed elsewhere.

Grazing. Until 1984 it was generally assumed that, because grass provided continuous soil cover and captured nitrate efficiently, leaching of nitrate from grassland was simply not a problem. That perception changed abruptly in 1984 with the publication of a paper[29] that drew attention to the consequences of the irregular deposition of urine and faeces by cattle, sheep, and other farm animals. The underlying problem is that these ruminant animals are not efficient converters of nitrogen into the products for which they are kept. For every 100 kg of nitrogen applied as fertilizer, as little as 10 kg may be recovered in meat, milk or wool. Much of the rest, about 80% of the nitrogen consumed by the animal, is excreted, and this is the source of most of the nitrate leached from grassland.

A cow urinates about $2 \, dm^3$ at a time on an area of about $0.4 \, m^2$, and this represents a nitrogen application of $400–1200 \, kg \, ha^{-1}$. This is far more than the grass can use. The safe application of $400 \, kg \, ha^{-1}$ that was mentioned above was an annual amount that would be split into several aliquots to ensure even growth throughout the season. This problem is exacerbated by the gregarious habits of cows and by the preference they often show for urinating in a particular part of the field, which may mean that only 15–20% of the area is 'treated' in this way. Returns of nitrogen in dung are smaller, only about half those of urine, but they are spread on an even smaller area, perhaps 7–10% of the field.

To assess the impact of nitrogen fertilizer in this system, we again need to consider how the efficiency of use changes as more nitrogen is applied. This involves the stocking rate, the number of animals per hectare. Using more nitrogen enables more animals to be kept, up to a maximum of six to ten per hectare,[30] and the productivity per hectare increases accordingly. Looking at the productivity per animal, however, shows a sharp decrease as the stocking rate increases. More nitrogen thus seems to imply not only more animals excreting but more excretion per animal. Measurements of nitrate losses from grazed grassland,[31] therefore, show a much larger increase in nitrate loss from the soil when the fertilizer application is increased from 200 to $400 \, kg \, ha^{-1}$ than when it is increased from 0 to $200 \, kg \, ha^{-1}$ (Figure 6). This experiment also investigated the effects of installing an efficient field drainage system. Such a system is an improvement in that it alleviates the damage that can be done when heavy-footed cows trample a water-logged grass sward. Drainage, however, also improved the

[29] J. C. Ryden, P. R. Ball, and E. A. Garwood, *Nature (London)*, 1984, **311**, 50.

[30] P. F. J. van Burg, W. H. Prins, D. J. den Boer and W. J. Sluiman, *Proc. Fertilizer Soc.*, 1981, **199**, 1.

[31] D. Scholefield, K. C. Tyson, E. A. Garwood, A. C. Armstrong, J. Hawkins and A. C. Stone, *J. Soil Sci.*, 1993, **44**, 601.

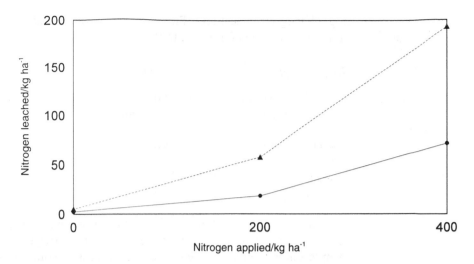

Figure 6 Losses of nitrogen as nitrate from grazed grassland related to the amount of fertilizer applied. The solid and broken lines refer to undrained and drained land, respectively. (Taken from Scholefield *et al.*[31])

conditions for the microbes in the soil, encouraging them to produce nitrate, and this and the freer movement of water through the soil resulted in a three-fold increase in the loss of nitrate from the soil (Figure 6). Other factors, such as the age of the sward and the time the animals are taken off the land, affect the issue, but experiments broadly suggest that 250 kg ha^{-1} is probably the most nitrogen that can be applied safely to grazed grassland in a year. This is not very different from the safe limit suggested for winter wheat.

Cutting the Grass and Feeding it Elsewhere. Grass is a safe, non-leaky crop, providing you keep cows off it. One option is therefore to grow grass using nitrogen fertilizer up to the safe limit (for grass without cows) of 400 kg ha^{-1}, and then make hay or silage and feed it to the cows somewhere where the urine and dung can be collected and used as manure. The resulting farmyard manure and slurry are a considerable resource; it has been estimated[2] that cattle in the UK excrete nitrogen worth about £300 million per year. These materials, however, are very much more difficult to apply than nitrogen fertilizer and, in particular, they are difficult to apply safely without risk of pollution of natural waters. They are not central to the theme of this review, which is about nitrogen fertilizer, but they are relevant to it because, if they could be used better, less nitrogen fertilizer would be needed. One practical problem that would need to be overcome is that most of the farmyard manure and slurry are produced in the west of the country, while the greatest demand for fertilizer is in the east.

3 The Contribution of Factors other than Nitrogen Fertilizers to the Nitrate Problem

Dividing the nitrate problem into topics is inevitably a somewhat arbitrary process, because of the interactions between the nitrogen fertilizer and the 'other factors'. The various crops, for example, could arguably have come in this section rather than Section 2, while the structure of the soil, discussed in this Section, might have gone in Section 2. Two other topics, the ploughing-up of old grassland and the deposition of nitrogen to land from the atmosphere clearly belong here.

Soil Structure

Soils differ greatly in the extent and manner in which they transmit water. Much depends on the proportions they contain for the four main particle categories:

- coarse sand   2.0–0.2 mm
- fine sand   0.2–0.02 mm
- silt   0.02–0.002 mm
- clay   < 0.002 mm.

The clay plays an important role in soil structure because its particles can adhere to each other strongly. This is a mixed blessing. Aggregation of soil particles to form 'crumbs' is vital for the proper functioning of some soils and this process depends on clay and organic matter to provide the adhesion and stabilization of the particles. On the other hand, a soil that contains a large proportion of clay may be so strongly held together that it lets very little water through except in the layer that has been ploughed. We can obtain a picture of the transmission of water and nitrate through soils by considering the behaviour of three general types: sandy soils, aggregated soils, and heavy clay soils.

Sandy Soils. The particles in sandy soils are relatively large, with correspondingly large spaces between them. Because these soils are also fairly homogeneous, water moves freely through much of the soil matrix. Any nitrate that is in the soil, whether from fertilizer or from microbial activity, is likely to be carried through the soil slowly but surely with little impediment. A sandy soil above an aquifer is usually seen as a threat to the quality of the water in the aquifer.

Aggregated Soils. Water takes the easiest path through the soil and so tends to flow round rather than through aggregates. This idea is often simplified to the concept that there are two categories of water in the soil, mobile and immobile, of which only the former moves through the soil. Nitrate that is immobile within aggregates is safe from leaching until it diffuses out into the mobile water in response to the gradient of concentration between the intra-aggregate pores and the mobile water outside. The leaching of the nitrate that is produced within aggregates by microbes is thus delayed by the aggregation of the soil. Fertilizer nitrate applied to a moist soil surface will tend to diffuse into the bulk of the soil where it will be relatively safe, but if an appreciable amount of rain falls before this happens, the nitrate will be carried by the rain into the mobile water in the soil, where it will pass between the aggregates. If the fertilizer nitrate is washed into, but not right through, the soil, some of it will diffuse into the safety of the aggregates, but heavy rain will carry some of it right through the soil.

Heavy Clay Soils. Heavy clay soils show an extreme form of the behaviour of water and nitrate in aggregated soils. Water cannot move through the matrix of such soils, except when it is imbibed by the dry soil. However, many of these soils shrink when dry and swell when wet, and therefore have cracks in them. These, together with wormholes and similar channels, form preferential pathways through which practically all the flow of water occurs. Because water flows through a very small proportion of the soil volume, it can move very rapidly, carrying nitrate with it. Many heavy clay soils must have artificial drainage[19] because they are so impermeable below 0.6–0.8 m. These drainage systems carry the water into ditches and other surface waters. If the preferential pathways connect with the drainage system, there is a very effective route along which the rain falling on the soil surface can carry nitrogen fertilizer down through the soil and into the ditch. These heavy soils often grow winter wheat, because they cannot be cultivated in early spring. This crop–soil combination can give very small leakages of nitrate in good circumstances, but much depends on the amount of rain that falls in the period following the application of nitrogen fertilizer.[20]

Ploughing-up Old Grassland

During this century, much old permanent grassland has been ploughed up to make way for arable crops, particularly cereals. This process began in the 1930s and accelerated during World War II, but it continued after the war and through the 1960s. By 1982 the area of permanent grass was one-third of what it had been in 1942. Enlarging the area of arable land may have increased the use of nitrogen fertilizer, but the reason the ploughing-up is relevant to this review is that at least part of the increase in the concentrations of nitrate in natural waters that has been blamed on nitrogen fertilizer may have resulted from the ploughing-up.

Some old permanent grassland at Rothamsted was ploughed in December 1959 and lost about 4 t ha^{-1} of organic nitrogen as a result of the activities of microbes and other soil organisms during the next 20 years. Much of this nitrogen was found as nitrate in the water in the chalk below, when a team from the Water Research Centre took cores from the chalk. These data, together with information on losses of organic nitrogen from grassland ploughed up in Lincolnshire, were used[32] to estimate, through a simple model, the potential contribution of the ploughing to nitrate concentrations in drainage from soils in England and Wales. The results were expressed in maps such as that in Figure 7, which shows its contribution to nitrate concentrations in various parts of the country in 1945. If the EC nitrate limit of 50 g m^{-3} had been in force then, it would have been disobeyed widely.

Could the ploughing of grassland in, say, 1945 really contribute to the nitrate problem in 1995? For this to be likely, we would need to be looking at some very long-term processes. Nitrate moves very slowly in some of the rocks that form our aquifers, at 0.8 m per year in unsaturated chalk, for example,[33] so nitrate entering the top of a chalk aquifer in 1945 will have just arrived at a water surface 40 m below in 1995. Forty metres is a fairly likely depth to the water surface, so this slow transport could mean that nitrate from the ploughing-up is still contributing to nitrate concentrations in deep aquifers. However, is the ploughing-up making an appreciable contribution to nitrate leaving the soil now?

[32] A. P. Whitmore, N. J. Bradbury and P. A. Johnston, *Agric. Ecosys. Environ.*, 1992, **39**, 221.
[33] C. P. Young, D. B. Oakes and W. B. Wilkinson, *Groundwater*, 1976, **14**, 426.

Figure 7 The estimated
contribution of nitrate from
ploughed-up old grassland
to concentrations of
nitrate-nitrogen leaving soils
in England and Wales in
1945. The EC limit is
11.3 g m^{-3} of
nitrate-nitrogen. (Taken
from Whitmore *et al.*[32])

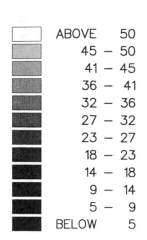

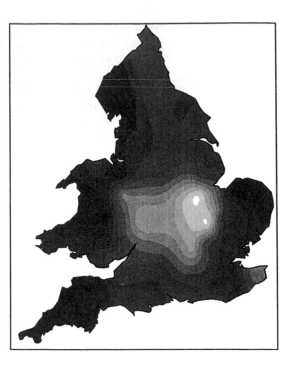

One of the relationships used[32] to obtain Figure 7 was that between the amount of organic nitrogen in the soil, N_{org} kg ha^{-1}, and the time, t yr, from the ploughing out of the permanent grassland. This was an exponential relationship derived by fitting to field data.[32]

$$N_{org} = 6733 + 3954 \exp(-0.132t) \qquad (2)$$

This relationship was used here to estimate what contribution the ploughing-up of old grassland in 1945 and at subsequent ten-year intervals could be making to nitrate-N in soil now in 1995 and, if it were leached, what contribution it then would make to the nitrate concentration, assuming 250 mm of through drainage per year (Table 3). The contribution of the 1945 ploughing is almost negligible, but that in 1965, if leached, would contribute more than one-third of the nitrate needed to reach the EC's limit. Any ploughing done in 1973 would contribute all the nitrate needed to reach the limit. If these calculations are correct, the war-time ploughing is now a spent force, but the considerable area of grassland ploughed up in the 1960s is still making some contribution to the nitrate in and leaching from the soil.

Nitrate Deposited from the Air

Rainfall at Rothamsted was first analysed for nitrate and ammonium about 120 years ago, but not very much was found. Between 1877 and 1915 the total deposition of nitrogen was 227 kg ha^{-1}, about 6 kg ha^{-1} per year on average. By 1990, measurements at four sites in south-east England[34] showed that the annual

Table 3 Effects of ploughing out of old permanent grassland at various dates. Estimated contribution in 1995 to trate-nitrogen in soil and, if leached, to the nitrate concentration in drainage from the soil (assumed to be 250 mm per year)

Year of ploughing	Contribution in 1995 to	
	Nitrate-N in soil (kg ha^{-1})	Nitrate concentration in drainage (g m^{-3})
1945	1	1
1955	3	5
1965	10	18
1975	37	66
1985	139	247

deposition had increased to 35—40 kg ha^{-1}. This, however, was a more detailed estimate that took account not only of nitrate and ammonium in rainfall but also deposition of nitrate on particulate matter and dry deposition of nitrogen oxides, nitric acid and ammonia. A more recent estimate[35] suggests that, at Rothamsted, about 37 kg ha^{-1} of nitrogen is deposited annually on bare soil and about 48 kg ha^{-1} on the soil with a well-established crop of winter wheat, with the difference reflecting the much larger surface area for deposition offered by the crop. The amount deposited when the wheat was grown is one quarter of the average application of nitrogen fertilizer to winter wheat in England and Wales. It comes from industry and motor traffic, each of which generates about half the nitrogen oxides, and from farming, usually nearby, which is responsible for most of the ammonia.[35] It is difficult to be certain what happens to the nitrogen that is deposited, but one estimate[36] suggested that about half was taken up by the crop and about 30% leached. Some runs with a currently unpublished model showed that one-quarter of the nitrogen deposited was leached and that it contributed about 15% of the nitrate leached overall. It may thus contribute 10–15 kg ha^{-1} to annual nitrate losses from soil. If we again assume 250 mm of drainage, its contribution to the nitrate concentration leaving is about 20 g m^{-3}, or 40% of the EC's limit, which leaves the farmer little room for error.

4 What Can Be Said About the Long-term Effects of Nitrogen Fertilizer?

In the Broadbalk Experiment at Rothamsted, winter wheat has been grown for the past 152 years. Some plots have received nitrogen fertilizer for the whole of this period, and any long-term harmful effects should surely have appeared by now. Crop yields are far greater than they were at the beginning, although this is attributable more to improved varieties than to any special properties of nitrogen fertilizer. There has been no discernible deterioration in the physical properties of the soil. Organic matter plays an important part in the physical stability of the soil, and is generally a useful indicator of its well-being. Its percentage in the soil

[34] K. W. T. Goulding, *Soil Use Mgmt.*, 1990, **6**, 61.

[35] *Impacts of Nitrogen Deposition on Terrestrial Ecosystems*, Report of the United Kingdom Review Group on Impacts of Atmospheric Nitrogen, Department of the Environment, London, 1994.

[36] Report of the AFRC Institute of Arable Crops Research for 1991, AFRC, London, 1992, p. 36.

increases regularly and statistically significantly as the annual application of nitrogen goes from 0 to 144 kg ha^{-1}. Within this range, more nitrogen means more crop and, therefore, more organic crop residues left in the soil. The microbes in the soil also seem to be in good health. The release of mineral nitrogen from organic matter is nearly twice as great in the plot given 144 kg ha^{-1} of nitrogen as in that given none. Ten of the plots have drains, and nitrate concentrations in the drainage are usually less than 50 g m^{-3}, although larger concentrations occur in some years at the start of drainage in autumn and shortly after the application of nitrogen fertilizer.[37]

What has Happened to Nitrogen Applications that Exceeded the Uptake by the Crop?

Some calculations made by the Agricultural Development and Advisory Service[38] suggest that up to 1976 farmers in England and Wales applied no more nitrogen to winter wheat than was removed in the crop. From 1977 onwards, however, more nitrogen was applied than the crop removed, and by 1986 the cumulative excess was more than 300 kg ha^{-1} of nitrogen. It would be interesting to know what happened to that nitrogen, and the author has attempted an assessment based on the following simple rules.

1. The fate of the nitrogen fertilizer applied in spring was assumed by harvest to be similar to the fate of the ^{15}N in the Rothamsted experiments discussed earlier.[13] On average: 65% in crop, 18% in soil organic matter, 1% as mineral nitrogen in soil, 10% denitrified, and 6% leached.
2. The excess was assumed to be made up by all the nitrogen that had not gone in the crop, and was thus 35%. Dividing the 35% into its components gave: 51% in the soil organic matter, 29% denitrified, and 20% leached. (The leached component included the 1% of the original application which was in the soil as mineral nitrogen at harvest.)
3. The fertilizer nitrogen in the soil organic matter was assumed to be remineralized in subsequent years according to the following pattern: 10% in the first year, 3% of the remainder in the second year, and 1% of the remainder in each of all subsequent years.
4. One-third of the nitrogen that was remineralized was assumed to be leached, while the remainder was taken up by the crop.

The excesses of nitrogen application over crop uptake in the individual years from 1977 to 1986 were read from Figure 4 of Sylvester-Bradley *et al.*[38] and subjected to the rules. Neither the leaching nor the denitrification losses seemed particularly large (Table 4), given that these were aggregate values for ten years, and the amount of nitrogen that was remineralized and then leached seemed very unlikely to be important.

If these calculations are right, 49% of the excess nitrogen was removed from

[37] K. W. T. Goulding and C. P. Webster, *Aspects Appl. Biol.*, 1992, **30**, 63.
[38] R. Sylvester-Bradley, T. M. Addiscott, L. V. Vaidyanathan, A. W. A. Murray and A. P. Whitmore, *Proc. Fertilizer Soc.*, 1987, **263**, 1.

Fate	Amount of N (kg ha^{-1})*	Total excess (%)
Leached during year of application	65	20
Denitrified during year of application	95	29
Remineralized from soil organic matter, then leached	7	2
Remineralized and taken up by crop	13	4
In organic matter at end	146	45
Total	326	100

*All the amounts shown, except that in the organic matter at the end, are totals for the 10 years.

the soil during the year of application and 45% of it was in the soil organic matter at the end of the period. A small annual release of mineral nitrogen from it will continue indefinitely. The excess that built up between 1977 and 1986 does not, therefore, seem to be a major problem, but we need to be aware that this excess has more than doubled since 1986 and the annual release with it. It is possible that 250–300 kg ha^{-1} has been added to organic nitrogen in soils since 1976, but is this a bad thing overall? We can view it as a potential long-term contributor to the nitrate problem, or we can take the old-fashioned approach and welcome it as a build-up in organic matter and soil fertility. It is, after all, less than 30 years since the UK was in a state of grave concern, not about the nitrate problem, but about declining amounts of organic matter in our soils. Remember the Strutt Report?[39]

Modelling the Long-term Effects of Nitrogen Fertilizer

Nitrate has been described as a 'biological time bomb'. Those who enjoy their 'doom and gloom' will be pleased to hear that there may be two such bombs, one with a physical clock and one with a biological clock. Both can be studied with the aid of computer models.

The physically timed bomb comprises nitrate that is moving slowly but inexorably through the unsaturated zones of chalk and other slow-response aquifers, as described earlier. This nitrate will eventually reach the water from which supplies are drawn for public use. This is a very long-term process, but it has been simulated quite successfully using a hydrological model[33] that makes allowance for mobile and immobile categories of water in the chalk similar in principle to those identified in many soils. The ploughing-up of grass is a key component in this model.

If there is a biologically timed bomb, it must lie in the accumulation in the soil's organic matter of nitrogen that will be released as nitrate by microbes at some unknown time in the future. The calculations made in the previous section suggest that, when nitrogen fertilizer is given in excess of the needs of the crop, a

[39] Ministry of Agriculture, Fisheries and Food, *Modern Farming and the Soil*, Report of the Agricultural Advisory Council on Soil Structure and Soil Fertility, HMSO, London, 1970.

substantial proportion of it remains in the organic matter as part of the potential problem. The turnover of organic matter during long periods, a century or more, has been simulated successfully[40,41] and variants of the models used have been applied to the turnover of nitrogen through organic matter. The SUNDIAL nitrogen model[42] was developed from the Rothamsted model for organic matter,[40] and is used with a time-step of one week to simulate the carry-forward of nitrogenous residues from one year to the next and their breakdown to ammonium and thence nitrate by microbes in the soil. The model has been used successfully to simulate the fate of nitrogen applied to one crop during two following crops. SUNDIAL is also being adapted for practical use as a source of advice for farmers planning fertilizer applications and for policy-makers concerned with the nitrate problem.

5 Conclusions

The Science of the Nitrate Problem

'All chemicals are dangerous'. So began the second leader in a well-known newspaper some years ago. Quite correct, and the most widely used laboratory chemical is the most dangerous; people drown in it every year. It has proved remarkably difficult to prove that nitrate is dangerous. So far as methaemoglobinaemia is concerned, there is no question that it can occur, but it is very rare and even more rarely fatal. When it does occur, it always involves water from wells rather than from a mains supply. We need to be aware of the possibility of sub-clinical cases at nitrate concentrations between 50 and $100\,\mathrm{g\,m^{-3}}$, but the risk to infants living where there is a mains water supply seems minimal. Stomach cancer was linked to nitrate in water in a very vigorous campaign in the early 1980s, but three carefully conducted epidemiological studies all failed to show any evidence for such a link. All that is clear is that nitrate is a hazard to the environment.

When we turn to the source of the nitrate problem, science is no kinder to popular conceptions than it was in respect of the supposed health risks. The general supposition that nitrate losses to natural waters sprang directly from the application of nitrogen fertilizer has been shown to be true to only a very limited extent. A much larger part of the problem in arable land seems to spring from what would otherwise be termed the natural fertility of the soil, the ability of soil organisms to make available nitrogen previously locked up in the soil's organic matter. Fertilizer from the bag can be timed to meet the needs of the crop and avoid the risk of leaching (as far as is possible). Naturally produced nitrate is timed by natural conditions and a substantial proportion of it is released at what has been shown clearly to be the least appropriate time. On grassland too, naturally produced nitrate seems to be the main problem, although in this case the internal workings of the cow, or sheep, are as important as those of soil organisms. What does seem clear, however, is that with both arable and grassland

[40] D. S. Jenkinson and J. H. Rayner, *Soil Sci.*, 1977, **123**, 298.

[41] W. J. Parton, D. S. Schimel, C. V. Cole and D. S. Ojima, *Soil Sci. Soc. Am. J.*, 1987, **51**, 1173.

[42] N. J. Bradbury, A. P. Whitmore, P. B. S. Hart and D. S. Jenkinson, *J. Agric. Sci.*, 1993, **121**, 363.

systems the farmer needs to think very hard before applying much more than $220–250\,kg\,ha^{-1}$ of nitrogen fertilizer.

The Politics of the Nitrate Problem

The farmer has inevitably been blamed for most of the nitrate problem, but we should not forget the contribution of the politician. The Treaty of Rome, signed in 1957, committed the six founder members of the European Community to the Common Agricultural Policy, and thence to the expansion of agricultural production and productivity through the provision of a guaranteed market for produce. This brought many social and economic benefits, but it led inevitably to the intensification of agriculture, and increased use of nitrogen fertilizer was a necessary part of the intensification package. When nitrate concentrations increased in natural waters, the EC introduced the $50\,g\,m^{-3}$ limit for nitrate in potable water in 1980, thereby putting the onus for solving the problem on farmers and water suppliers. It should perhaps have reviewed the Common Agricultural Policy at the same time.

The UK became a member of the Community in 1973 and came under the Common Agricultural Policy as a consequence. This had a dual effect on the use of nitrogen fertilizer. Better prices for crops made it economic to use more nitrogen. Also, as farming became more profitable, land became more attractive as an investment, particularly to large institutions such as pension funds. Land prices therefore rose about five-fold between 1970 and 1980 (but by much less when corrected for inflation), and interest rates also rose, making it essential to maximize the economic return on the crop. This could be done most effectively by using more nitrogen fertilizer. The overall result was that the average application of nitrogen fertilizer, taking all arable crops and grass together, increased from about $80\,kg\,ha^{-1}$ in 1970 to about $140\,kg\,ha^{-1}$ in 1984. It is perhaps worth recalling that it was in 1977 that applications of nitrogen to winter wheat first began to exceed what was removed by the crop.

These changes obviously did not come about just because of the Common Agricultural Policy. Some would have happened in any case, and the British Government must have contributed to them when it stated in a White Paper in 1975 that it took the view that 'a continuing expansion of food production in Britain will be in the national interest'. The demands of the public also need to be remembered. The demand for plentiful cheap food could only be met through intensification of agriculture. This inevitably clashed with the demand for supplies of water that were not only fit for drinking but available in sufficiently vast quantities to wash the nation's cars and sprinkle the nation's gardens.

The nitrate problem is sufficiently political in its origins to need a political solution. The Common Agricultural Policy is under reform, giving opportunities for new initiatives. Much will depend on how the European Community responds to the conflicting objectives of liberalizing trade, meeting environmental standards and maintaining food supplies. If the right balance can be struck, one of the benefits should be an amelioration of the nitrate problem. Three aspects seem important: effective transfer of information and technology from scientists to users, decoupling support for environmental objectives from the management of production and long-term strategic planning that integrates policies for the supply of food with those for the supply of water. Finally, we need to keep in mind Havelock Ellis's description of progress as 'the exchange of one nuisance for another nuisance' and make sure that any solution we find to the nitrate problem does not simply replace it with another, equally intractable, environmental problem.

Eutrophication of Natural Waters and Toxic Algal Blooms

ALASTAIR J. D. FERGUSON, MICK J. PEARSON AND COLIN S. REYNOLDS

1 Introduction

The term eutrophication is usually understood to refer to the excessive primary production (usually of algae but higher plants, or water weeds, are not excluded) in water, caused by an enhanced supply of nutrients. The process has been recognized since the early 1920s,[1] although the appreciation of the relationships between the nutrient supply, the basin morphology, the hydrology, the hydrography, and the fertility of a lake basin is as old as limnology itself. However, an expanding human population, with increasing aspirations of living and sanitation standards, and an accompanying increase in the intensity of agricultural production, have led, in many locations, to severe enrichment of waters and a visible deterioration of water quality. These symptoms have been most apparent in Europe and North America, but they are now common in many other parts of the world.[2]

The major concerns about eutrophication include: problems with the treatment of water for potability, where algae and their chemical products have to be removed; deterioration in freshwater commercial and sport fisheries, through choking weed growth and poor transparency; deoxygenation of lakes and rivers, as microbial communities break down the primary products of plant growth; obstruction of water flow in rivers by dense stands of plants; and the reduction in the diversity of animal and plant populations (or, at least, the loss of the more desirable species). Mostly, however, it is the unsightliness of blooms and surface scums of algae on lakes, reservoirs, estuaries and coastal waters, combined with the increasing frequency of the reported incidences of toxicity of the bloom-forming organisms themselves, which do most to give eutrophication its bad name.[3]

Public concern about the abundance of algae, and of the toxic cyanobacteria in particular, was raised by events in the UK in the summer of 1989 which involved the deaths of dogs and sheep at Rutland Water, Leicestershire, and the acute

---

[1] H. Bernhardt, in *Eutrophication: Research and Application to Water Supply*, ed. D. W. Sutcliffe and J. G. Jones, Freshwater Biological Association, Ambleside, 1992, p. 1.

[2] D. M. Harper, *Eutrophication of Freshwaters*, Chapman and Hall, London, 1992.

[3] C. S. Reynolds, in *Phosphorus in the Environment: its Chemistry and Biochemistry*, ed. R. Porter and D. Fitzsimmons, Excerpta Medica, Amsterdam, 1978, p. 210.

poisoning of soldiers who had been swimming in Rudyard Lake, Staffordshire.[4] Since that time, further well-publicized events involving blooms of toxic cyanobacteria, at Loch Leven, Scotland, and along the Murray-Darling River in Australia, have promoted renewed interest in toxic production, adverse human health effects, and in the possible ways to control the development of populations of the cyanobacteria.

The management of problems related to toxic cyanobacterial blooms in the UK has been met in the short term by improving public awareness of the danger of contact with potentially toxic scums, by establishing and clarifying the responsibilities of public bodies in dealing with incidents of bloom development, and by introducing comprehensive monitoring programmes.

In the longer term, the approach to the management of water quality lies in the development of sound strategies to deal with the sources of nutrients within entire drainage basins. In turn, this requires the integration of knowledge and understanding of the relevant processes taking place within catchments, in order to better diagnose and prescribe appropriate strategies for improving and maintaining the aquatic ecosystems. In the medium term, responsive tactics will doubtless continue to be applied; even here, there is a pressing need to recognize the choice of local management therapies that are available and to be able to select the option most likely to yield success against the investment made.

This review seeks to identify the key areas in aquatic science that are required to guide the future management of algal problems in freshwaters. Although the factors driving primary production have challenged scientific research for many years, questions remain about the scientific approaches needed to pull together the appropriate practical guidance for management. Given the theme of this volume, however, special attention is given to the role and influence of agricultural chemicals.

2   The Role of Agricultural Fertilizers in Aquatic Production

Agricultural fertilizers are generally considered to play a major role in eutrophication. This is evident in the case of the very soluble salts of nitrogen, which are readily leached by drainage into watercourses. However, there is doubt about the biological activity of phosphorus released, other than as in the soluble and reactive orthophosphate fractions, mostly the ions HPO_4^{-2} and $H_2PO_4^{-}$, because these other forms of phosphorus are not readily available to plant uptake. Before discussing the mechanisms in more detail, it is essential first to address the general relationships governing the assembly (production) of plant and animal biomass. However, the influence of nutrients derived from agriculture cannot be described or quantified without considering other, and perhaps more important, influences on primary production.

There is a general understanding of the reasons why nutrients are critical to the productive capacity of biological systems. The dry biomass of plants and animals comprises some 20 elements, the predominant atoms being those of carbon, hydrogen, oxygen, and nitrogen. Moreover, ideally they are required in fairly

---

[4]   National Rivers Authority, *Toxic Blue-green Algae*, Water Quality Series Report No. 2, NRA, 1990, p. 128.

inflexible molecular ratios. Those which are more scarce, or chemically less-readily available relative to the requirements of the *autotrophic* (or self-nourishing) plant, are clearly the most likely to fall into short supply and thus to regulate, or *limit*, the assembly of new biomass. They are therefore referred to as 'limiting factors' or 'limiting nutrients', although confusion persists about the sense in which the term is often applied, *i.e.* to refer to the capacity of a system to support biomass, or to the rate of assembly at which the capacity can be approached.[5]

Populations of planktonic *photoautotrophs* (light-utilizing autotrophs) are just as vulnerable as any other group of plants to a limiting flux of one or other of these nutrients. Conversely, raising the rate of supply of these nutrients is just as likely to promote growth and to lead to levels of plant biomass that adversely affect the use of the lake or watercourse.

The proximal sources of nutrients in lakes and rivers include rainfall, the drainage area upon which it falls and the extent to which it is modified by its contact with the ground. Geography, geology, geomorphology and hydrology interact to determine the time the water stays on or in the ground and the time available for contact with the appropriate solutes. The terrestrial vegetation is seeking the same elements, which often results in a further deficiency, relative to demand, in the concentration of those elements in the drainage water. Among freshwaters generally, nitrogen and phosphorus are often received in low concentrations in the drainage water from the land. This is especially true of old landmasses clothed with established forest. The main chemical composition of run-off from forests to receiving waters is a dilute solution of non-growth-limiting elements and a complex range of organic carbon products.

A recent review[6] of research on phosphorus input to surface waters from agriculture highlights the variability of particulate and dissolved phosphorus contributions to catchments. The input varies with rainfall, fertilizer application rates, the history of the application of the fertilizer, land use, soil type, and between surface and sub-surface water. The balance struck between export of nutrients from the catchment and recipient-water productivity is the primary factor which controls its quality.

Anthropogenic effects in the same catchments, including those emanating from human settlement, clearance of forest for agriculture and the development of mineral extraction and manufacturing industry, clearly alter the contemporaneous balance and so change the supply of nutrients to the water. The extent of this effect is no less clearly related to the scale of human activity as it is to the hydrological throughput from the catchment. A summary of the order of export rates of nitrogen and phosphorus per unit area from different categories of land use, is given in Tables 1 and 2.

The loads of nitrogen and phosphorus to receiving waters in the UK have increased greatly in the last 30–40 years. Even Windermere, England's largest natural lake, received twice the annual load of nitrogen in 1991 than it received in the 1950s and just over three times more phosphorus. Of greater significance,

[5] C. S. Reynolds, in *Eutrophication: Research and Application to Water Supply*, ed. D. W. Sutcliffe and J. G. Jones, Freshwater Biological Association, Ambleside, 1992, p. 4.

[6] R. H. Foy and P. J. A. Withers, *The Contribution of Agricultural Phosphorus to Eutrophication*, The Fertiliser Society, Peterborough, 1995.

Table 1 Total phosphorus (TP) losses from different types of catchment*

Catchment type	TP	Reference
Undisturbed temperate forest	2	Hobbie and Likens, 1973[7]
Undisturbed boreal forest	3–9	Ahl, 1975[8]
Cleared forest, igneous catchment	5 ⎫	Dillon and Kirchmer, 1974[9]
Cleared forest, sedimentary	11 ⎬	
Cleared forest, volcanic	72 ⎭	
Pasture (US)	8–20	Johnston *et al.*, 1965[10]
Intensively arable (Netherlands)	25	Kolenbrander, 1972[11]
Intensively arable (UK)	7–25	Cooke and Williams, 1973[12]
Intensively arable (with heavy erosion, Missouri)	190	Schuman *et al.*, 1965[13]
Intensively arable (with waterlogging, California)	40–53	Johnston *et al.*, 1965[10]
Mixed upland (Windermere)	34	Atkinson *et al.*, 1985[14]
Groundwater leachate (Crose Mere)	35–93	Reynolds, 1979[15]
Urban runoff	~100	Harper, 1992[2]

*In $kg\,P\,km^2\,yr^{-1}$; mostly from Harper (1992).[2]

Table 2 Total nitrogen (TN) losses from different types of catchment*

Catchment type	TN	Reference
Temperate forest	~200	Hobbie and Likens, 1973[7]
Boreal forest	90–60	Ahl, 1975[8]
Pastures, low intensity	100–1000	Ahl, 1975[8], Gachier and Furrer, 1972[16]
Pastures, high intensity	2000–25 000	Various, see Harper, 1992[2]
Arable, cereals	4000–6000	Various, see Harper, 1992[2]
Arable, cashcrops	<10 000	Stewart *et al.*, 1973[17]
Mixed pasture and arable	2500–3000	Troake *et al.*, 1973[18]
Mixed upland (Windermere)	530–638	Atkinson *et al.*, 1985[14]
Groundwater leachate (Crose Mere)	427–638	Reynolds, 1979[15]
Urban runoff	~1000	Harper, 1992[2]

*In $kg\,N\,km^2\,yr^{-1}$; mostly from Harper (1992).[2]

[7] J. E. Hobbie and G. E. Likens, *Limnol. Oceanogr.*, 1973, **18**, 734.

[8] T. Ahl, *Verh. Int. Verein. Theor. Angew. Limnol.*, 1975, **19**, 1125.

[9] P. J. Dillon and W. B. Kirchmer, *Water Res.*, 1974, **9**, 135.

[10] W. R. Johnston, F. Ittihadiel, R. M. Daum and A. F. Pillsbury, *Proc. Soil Sci. Soc. Am.*, 1965, **29**, 287.

[11] G. J. Kolenbrander, *Stikstaff*, 1972, **15**, 56.

[12] G. W. Cooke and R. J. B. Williams, *Water Res.*, 1973, 7, 19.

[13] G. E. Schuman, R. E. Burwell, R. F. Priest and M. D. Schuldt, *J. Environ. Qual.*, 1973, **2**, 299.

[14] K. M. Atkinson, J. M. Elliott, D. G. George, D. C. Jones, E. Y. Howarth, S. I. Heaney, C. A. Mills, C. S. Reynolds and J. F. Talling, *A General Assessment of Environmental and Biological Features of Windermere and their Susceptibility to Change*, Freshwater Biological Association, Ambleside, 1985.

[15] C. S. Reynolds, *Field Stud.*, 1979, **5**, 93.

[16] R. Gachier and D. J. Furrer, *Schweizer. Z. Hydrol.*, 1972, **34**, 41.

[17] W. P. D. Stewart, E. May and S. B. Tuckwell, *Tech. Bull., MAFF*, 1973, **32**, 276.

[18] R. P. Troake, L. E. Troake and D. E. Walling, *Tech. Bull., MAFF*, 1973, **32**, 340.

however, is the fact that almost all of the increase arrived in solution, meaning that the readily available nutrient fraction increased over the same period by almost 14-fold. Against the background of palaeolimnological studies carried out on this lake, it was also possible to judge that there had been a slow enrichment of the lake following the Iron Age and later forest clearance, pastoral farming, human settlement and quarrying. The change since 1950 was abrupt and unlike any period before it. It coincided with the greater use of fertilizers for enhanced grass production, with a greater resident population, an increase in the annual influx of tourists, the general introduction of piped water to homes, the introduction of sewerage and the installation of full (secondary) treatment of foul drainage.[19]

The coincidence of these events, generally, has made their precise effects difficult to separate. For the English Lakes, the effects have been largely attributable to the addition of phosphorus, but in eastern England, for example, where human population was already high, it has been the intensification of agriculture which has been primarily responsible for the enhanced availability of nitrogen. In neither case, however, is the mechanism of added biological production explained, nor is the approach to providing a solution readily identified. The view that 'added phosphorus comes from sewage and added nitrogen comes from agriculture' has some factual basis, but is a gross oversimplification. We must also consider the nature of the biological response of the waterbody to the nutrient added.

3 The Factors Controlling Algal Blooms

The simple deduction may be made that, if the export of nitrogen and, especially, of phosphorus is increased, there is some *pro rata* increase in the capacity of the water to support primary production which will, eventually, be fulfilled. This relationship is quantitatively expressed in equations.[20] These equations relate the known phytoplankton mass as some analogue (mean annual or mean summer chlorophyll content) to the phosphorus available (hydraulically and bathymetrically weighted annual load, or a mean in-lake concentration) in order to derive a general statement about the long-range behaviour of a series of lakes in respect to potential phosphorus availability. The regression is valid over five orders of magnitude in phosphorus availability. Analogous models for nitrogen loads[21,22] exist, but there is a less satisfying fit of biomass analogues over two orders of in-lake nitrogen concentrations. This fact compounds the view of Schindler,[23] and limnologists generally, that the chemical concentration most relevant to the capacity to support phytoplankton biomass is that of phosphorus.

The regressions average out a vast amount of within-site variability and between-site fluctuation about the mean. To ignore or suppress these regressions greatly reduces the potential opportunities to manage the algal populations.

[19] J. W. G. Lund, *Gidrobiol. Zh.*, 1978, **14**, 10.
[20] R. A. Vollenweider and J. Kerekes, *Prog. Water Technol.*, 1990, **12**, 5.
[21] M. Sakomoto, *Arch. Hydrobiol.*, 1966, **62**, 1.
[22] J. W. G. Lund, *Water Treatment Examin.*, 1970, **19**, 332.
[23] D. W. Schindler, *Science (Washington, DC)*, 1977, **196**, 260.

Production through much of the year will be subject to other constraints; for example, the availability of light beneath the water surface. Seasonal differences in day length and periodic fluctuations in the depth of light penetration by active wavelengths often have an overriding effect on the net production rates and the supportive capacity.

Temperature also affects production rates but, through its influence on the thermal expansion of water, it also induces changes in the depth of vertical mixing and resistance to wind-stirring processes. Reactions to temperature of other components of the food chain are also important in the regulation of phytoplankton biomass by consumers. Different phytoplankton species, with important morphological differences, are differentiated selectively by the interplay of these factors.[24]

In this way, the near-linear chlorophyll–phosphorus relationship in lakes depends upon the outcome of a large number of interactive processes occurring in each one of the component systems in the model. One of the most intriguing aspects of those components is that the chlorophyll models do not need to take account of the species composition of the phytoplankton in which chlorophyll is a constituent. The development of blooms of potentially toxic cyanobacteria is associated with eutrophication and phosphorus concentration, yet it is not apparent that the yield of cyanobacterial biomass requires any more mass-specific contribution from phosphorus. The explanation for this paradox is not well understood, but it is extremely important to understand that it is a matter of dynamics. The bloom-forming cyanobacteria are among the slowest-growing and most light-sensitive members of the phytoplankton.[25]

It is not surprising that, globally, cyanobacteria do not thrive at high latitudes and that, in the UK, the highest densities are found in the summer months. Even then, they have to exist in competition with metabolically more active smaller species which can quickly produce a much bigger proportion of the biomass, or they have to do so in the wake of cold-tolerant diatoms and chlorophyte algae that bloom in the spring months. The spring growth removes phosphorus from the water in commensurate quantities, much of which is then transported, with the species of spring-blooming algae, to the sediments. In many lakes, by the time that the physical growth conditions are suitable for bloom-forming cyanobacteria, should they be present, the chemical conditions can represent extreme nutrient deficiency. It is only when the nutrient supply is sufficient, or sufficiently sustained, to remain available after the spring bloom, that these cyanobacteria can expect to sustain growth and reach bloom proportions.

This is the principal linkage between cyanobacterial blooms and eutrophication. Avoidance of cyanobacterial production does not necessarily depend upon eliminating all phosphorus inputs, but upon ensuring that optimum physical and chemical conditions for these organisms do not coincide. It is easy to understand why the biggest blooms in the UK have been in fertile lakes and reservoirs after prolonged spells of warm, dry weather in summer.

[24] C. S. Reynolds, in *Plankton Ecology*, ed. U.Sommer, Brock Springer, Madison, 1989, p. 9.
[25] C. S. Reynolds, *Toxicity Assessment*, 1989, **4**, 229.

Table 3 The range of phosphate fractions according to their availability to algae

Most available					*Least available*
Dissolved phosphorus	Loose-bound phosphorus	$Fe(OH)_3$-bound phosphorus	Phosphate bound to metal oxides	Carbonate-bound phosphorus	Organic and refractory phosphorus
Water extractable	NH_4Cl extractable	Dithionite extractable	NaOH extractable	HCl extractable	Residual

4 Sources of Nutrients

The main diffuse sources of phosphorus and nitrogen affecting the fertility of lakes are shown in Tables 1 and 2. The forms in which phosphorus is supplied to recipient waters create a further complexity in understanding the significance of the loadings. The role of phosphorus in lake fertility was recognized by classical limnologists such as Pearsall, Birge and Juday, even before phosphorus could be measured accurately. The standard method adopted (the acid molybdate spectrophotometric method for determining orthophosphate phosphorus or soluble reactive phosphorus, SRP), however, is considered to be misleading, because the small quantities detected in water samples exclude phosphorus already stored in plankton cells. The initial digestion of whole water samples, in order to derive values for 'total' phosphorus (TP), has also become a standard method, but the danger is the opposite one of assuming that all this phosphorus is, or could ever become, available to algal cells. There is a wide range of different forms of phosphorus which differ in their availability for uptake by algae and the only way to separate them is to undertake a progressive, serial analysis.

The various forms of phosphorus are summarized in Table 3 and labelled according to the extractant necessary to leach them from the original material. The first two represent labile and substitutable phosphorus fractions in the water which are readily available to algae. The iron-bound fraction is often the largest fraction, but it is occluded within Fe^{3+} matrices and is not available for uptake until redox gradients reduce Fe^{3+} to Fe^{2+}. This is the form of phosphorus which comes into solution in the anoxic deep water of eutrophic lakes in summer. Phosphorus bound to oxides and clay minerals is also scarcely available to algae except at high pH or when they produce phosphatases. The last two forms shown in Table 3 are effectively unavailable. Thus, a portion of the TP estimation is never available to contemporary organisms and much of the remainder can be ignored from day-to-day budgets, but the redox- and pH-sensitive fractions remain part of the potential fertility of a given water.

Looking at these sources from an algal perspective, evolution in habitats in which soluble sources are deficient or, at best, transiently present in time as well as in space, has provided them with mechanisms to take up soluble phosphorus at rapid rates from low concentrations. Most can satisfy their growth needs at SRP concentrations well below 10^{-7} molar, even though maximum uptake rates are some 10–100 times the maximum rates of consumption by growth.[26] This, in turn, means that at SRP concentrations greater than 10^{-7} M ($\sim 3\,\mu$g P l^{-1}) it is safe to assume that algal growth rate cannot be phosphorus limited. Equally, phosphorus deficiency will be represented by nearly undetectable SRP levels.

It should be pointed out that, significantly, the main increment to the total phosphorus content of recently eutrophied waters is in the SRP fraction delivered by secondary sewage treatment. While certain types of animal husbandry and market gardening may imitate enrichment by sewage effluent, the main agricultural export of phosphorus is in particulate form: as clay, fine sand, and detritus. It represents a contribution with a potential to stimulate biological production, but its role in recent incidences of lake and river eutrophication is assumed to be subordinate to secondary sewage treatment. This may also be inferred from the rapid improvement in the condition of lakes benefiting from the tertiary treatment of sewage inputs (to remove most of its soluble phosphorus content). In such circumstances, diffuse inputs (often from agriculture) are unchanged. As has already been pointed out, modern agriculture also contributes to inflated loadings of nitrogen. As long as the productive capacity of waters is constrained by phosphorus, or any other physical or chemical factor, the additional nitrogen need not provoke exaggerated plant growth. On the other hand, the application of inorganic nitrogen fertilizers has contributed to changes in fertility and aquatic species composition in the small lakes ('meres') of Cheshire and Shropshire which have a high, apparently natural, supply of phosphorus.[15]

When considering the availability of nutrients, it is also necessary to examine the significance of nutrient re-use within the waterbody. These 'internal' sources amount not to an additional load, but a multiplier on the recyclability of the same load. This nutrient recycling and the internal stores from which they are recycled are often misunderstood, but there is a dearth of good published data about how these recycling mechanisms operate. Microbial decomposition in the water column is one of several internal loops recognized in recent years, but these are not closed and the flux of nutrients recycled through them is delayed rather than retained.

The destiny of most biological material produced in lakes is the permanent sediment. The question is how often its components can be re-used in new biomass formation before it becomes eventually buried in the deep sediments. Interestingly, much of the flux of phosphorus is held in iron(III) hydroxide matrices and its re-use depends upon reduction of the metal to the iron(II) form. The released phosphate is indeed biologically available to the organisms which make contact with it, so the significance attributed to solution events is understandable. It is not clear, however, just how well this phosphorus is used, for it generally remains isolated from the production sites in surface waters. Moreover, subsequent oxidation of the iron causes re-precipitation of the iron(III) hydroxide flocs, simultaneously scavenging much of the free phosphate. Curiously, deep lakes show almost no tendency to recycle phosphorus, whereas shallow

---

[26] C. S. Reynolds, in *Urban Waterside Regeneration*, ed. K. N. White, E. G. Bellinger, A. J. Saul, M. Symes and K. Hendry, Ellis Horwood, Chichester, 1993, p. 330.

lakes do so nearly continously.[27] There remains an urgent need to quantify and model these rapid, oxic cycles of biologically available phosphorus.

There is a further complication in shallow lakes containing macrophytes (aquatic flowering plants, pteridophytes, and macroalgae). These take up and accumulate nutrients from the water and from the aquatic 'soil' in which they are rooted (sediment). Although these plants are sometimes classed as nuisance weeds, they nevertheless act as an important alternative sink for nutrients which are denied to the plankton. In recent times, a key role of macrophytes in the successful and sustained management of water quality has been identified and explained.[28]

5 Control of Eutrophication and Toxic Blooms

Over the past 30 or more years there have been many attempts to control the symptoms of eutrophication, particularly in lakes. Most have sought to reverse the trends in nutrient enrichment and have experienced varying degrees of success. The initiatives have ranged from diversion of nutrient-rich water around lakes to complex combinations of nutrient removal from point sources in the catchment, nutrient reduction from diffuse sources and in-lake management of the biology and chemistry.[29,30] Generally the approach is a strategic and long-term one. If we recall the description of factors affecting the development and growth of algal blooms and other aquatic plant communities in Section 3, this showed that populations are unable to reach the potential light-limited maximum, without an adequate supply of nutrients. Equally, it also demonstrated that other factors such as light, especially with respect to its underwater availability to phytoplankton, can regulate its production and growth for long periods of the year. In fact, a combination of such factors, acting serially rather than simultaneously, drives the seasonal changes in mass and species composition in the planktonic community. This offers a choice of solutions for combating the symptoms of eutrophication.

The extent to which any of the mechanisms might predominate in a given receiving water should be the primary influence in the selection of control methods. Evaluation of the extent to which each factor already regulates production will quickly identify the most appropriate methods for managing the problems. Simple expert systems are already available to make the necessary comparisons (*e.g.* the National Rivers Authority's Prediction of Algal Community Growth And Production, PACGAP). Projected outcomes can also be weighed against costs of implementation of the various methods available, in order to test cost–benefit criteria.

The options themselves invoke different controls and philosophies. The main techniques may be categorized according to whether they deal with the nutrients directly, or whether they interfere with the physical growth environment, or whether they invoke some form of biological remediation.

---

[27] C. S. Reynolds, *Freshwater Biol.*, 1996, in press.

[28] M. Scheffer, S. H. Hosper, M.-L. Meijer, B. Moss and E. Jepperson, *Trends Ecol. Evolution*, 1993, **8**, 275.

[29] H. Sas, *Lake Restoration by Reduction of Nutrient Loading*, Academia Verlag Richarz, Skt Augustin, 1989, 498 pp.

[30] G. L. Phillips and R. Jackson, *Verh. Int. Verien. Theor. Angew. Limnol.*, 1996, in press.

Nutrient Removal

Intuition dictates that reducing nutrient inputs, particularly phosphorus, must reduce the supportable biomass. Relationships[29] show that sustained response cannot be effected before it can be demonstrated (i) that nutrient is exhausted at a lower concentration of algae than the existing maxima and (ii) that *in situ* recycling is unable to make up the shortfall.

Reducing the amounts of nitrogen generated by run-off from the land is of limited consequence to a predominantly phosphorus-regulated system. Even if brought to the level of limitation, nitrogen fixation by microorganisms, including species of cyanobacteria, is always likely to make up the difference to achieve the phosphorus-determined carrying capacity. If the control of phosphorus looks more promising in this respect, it should be established which fractions of the TP are immediately available and how these might be metabolized in processes operating within the receiving body. In order to regulate toxic cyanobacteria, it is sufficient to achieve removal of enough phosphorus to reduce the post-spring capacity to a low level.

As the most significant point sources of phosphorus are those from sewage treatment works (STW), control of phosphorus loading is most readily achieved either by precipitation of phosphorus with iron salts (iron(III) sulfate or iron(III) chloride) or by biological removal. The latter can only effectively be achieved in STWs using activated sludge and there have been many descriptions of this technique.[31]

Under optimum conditions of anaerobic and aerobic treatment in activated sludge plants, up to 90% removal of phosphorus can be achieved. Biological removal can, at best, achieve effluent concentrations of about $0.2\,mg\,l^{-1}$ total phosphorus but more usually reduces phosphorus to about $1\,mg\,l^{-1}$.

A problem of many sewage treatment works in the UK is that urban drainage is included with domestic sewage in the sewage collection systems. The resulting storm-water discharges, at times of heavy rainfall, lead to continuing phosphorus export to the river. Correction of this problem requires considerable investment in separate systems.

In the short to medium term, nutrient export to water might be better managed through the widespread adoption of agricultural best-practice. In particular, discharges from silage clamps, dairy yards and animal rearing facilities should be separated from farm drainage and treated appropriately. Artificial macrophytic wetlands can be used successfully to reduce nutrients from point source effluents in many of these situations. However, this particular practice has not been used in the UK. Most of the experience of the use of these artificial wetlands is in America and Australia. The efficacy of these methods depends very much on the design, particularly in relation to the hydraulic retention time, and the performance, which varies with time.[32] These systems can achieve a 50% reduction in phosphorus concentrations, and so are most appropriate to address smaller point source discharges.

[31] S. Williams and A. W. Wilson, *J. IWEM*, 1994, **8**, 664.
[32] A. D. Marble, *Guide to Wetland Function and Design*, A. D. Marble & Co., Rosemount, PA, 1990, p. 126.

Catchment Management

The longer term approach to managing eutrophication and its consequences should be based upon better coordination and management of land-use and planning policies of entire catchments of rivers, lakes, and reservoirs. The approaches already adopted for reducing diffuse sources of nitrogen have been employed for the protection of drinking water quality rather than eutrophication. The methods used have included crop zonation and reduced fertilizer application, which arguably represent more responsible and more sustainable land husbandry. Much better resolution of the nutrient exports summarized in Tables 1 and 2, and for more types of land-use activity, is required.

The techniques for reducing the input of nutrients from agriculture have been developing rapidly and include:

- reduction in soil erosion by better land management
- retention of nutrients by buffer zones
- retention of nutrients in ditches parallel to watercourses
- the use of artificial wetlands at the ends of feeder streams
- increasing instream nutrient metabolism by stream rehabilitation (re-instatement of riffle/pool sequences)

All of these techniques are available, but have not been well researched in terms of their nutrient removal efficiency. One exception is the recent work on the efficiency of buffer zones,[33] which used figures of 10–15% for nitrogen and 20–30% for phosphorus reduction by wooded buffer zones in a study of the Slapton Ley catchment.

In all of the techniques which use artificial barriers to surface run-off of nutrients there is a need to consider the influence of land drains. If these are widespread in a catchment a reduction in nitrogen loading to the watercourses will be unlikely, because the nitrogen is predominantly dissolved and runs through the sub-soil to the drains. Phosphorus control by these barriers will be less affected by land drains because the main input of the phosphorus is in the particulate form which would be prevented from running off the surface to the watercourses.

In addition to the use of techniques to reduce nutrient run-off at source, entry of nutrients to lakes can be reduced by using pre-lake techniques. These include the use of artificial wetlands on inflow streams, the use of iron salt treatment in pre-lake lagoons or by installing treatment plants to remove phosphorus.

Most lakes affected by eutrophication will also have significant amounts of phosphorus in their sediments, which can be recycled into the water column (Section 4). The control of this source can be achieved by treating the sediments with iron salts or calcite to bind the phosphorus more tightly into the sediments. These methods have been used to some effect, but consideration has to be given to the quality of the materials used and whether or not the lake can become de-oxygenated in the summer. In the latter case this can be overcome by artificial de-stratification.

In smaller shallow lakes, where the internal re-cycling of phosphorus can be

---

[33] H. M. Wilson, M. T. Gibson and P. E. O'Sullivan, *Aquat. Conserv.*, 1993, **3**, 239.

highly significant, removal of the nutrient-rich sediment can be an effective option. However, in many cases there are considerable practical problems with the removal of the sediment.

Physical Controls

The methods available include enhanced flushing, enhanced vertical mixing, altered mixing frequency, and food-web manipulation.

One attractive approach to the problem of a lake or reservoir receiving large nutrient loads from dispersed sources or from diffuse non-point sources, is to manage the hydrography of the system.[34] Provided that the waterbody is deep enough to stratify naturally, that the volume of bottom water is at least as great as the surface mixed layer and that light penetrates to less than half of the fully-mixed volume, destratification is likely to be effective.

Physical controls are generally only applicable in lakes. The influence of river morphology on eutrophication is not sufficiently well understood to be used effectively. The exception to this would be the short-term use of high flow to reduce the retention time to levels which limit growth rates of nuisance species such as cyanobacteria.

The most commonly used physical method for long-term eutrophication control in lakes is that of artificial destratification. This method is well tried and understood and uses either jetted water or compressed air bubbles to break down the lake stratification in the summer months. Algal growth is also affected by an increase in circulation. This is due to the artificial shading effect which results from the algae spending less time near the surface and consequently less time in the light. This technique also reduces the redox-dependent phosphorus release from sediments because the sediment surface remains aerobic.

In most cases, artificial destratification is carried out throughout the summer. However, work in the Lund Tubes in Blelham Tarn[35] suggests that intermittent destratification could provide further enhancement to this method of control by creating an unstable environment. This instability causes disruption of growth patterns of the algal community and, consequently, no species becomes dominant.

Artificial pumped storage reservoirs, such as those built to supply London, have been constructed to reduce problems with algae. These have a circular design, with pumped inflows which are jetted to mix fully the water column.[34] However, although these effectively deal with eutrophic water, they are not solutions which are normally used for rehabilitation of eutrophic lakes.

Aeration of the hypolimnion (lower, colder layer of water in a stratified lake) without disruption of stratification has been used in deep lakes. This has the advantage of not increasing the temperature of the hypolimnion and prevents the advection of nutrient-rich water into the epilimnion (upper, warmer layer of water in a stratified lake). Oxygen injection is preferred in order to prevent the build up of nitrogen super-saturation which is toxic to fish.[36]

[34] J. A. Steel, in *Conservation and Productivity of Natural Waters*, ed. R. W. Edwards and D. J. Garrod, Academic Press, London, 1972, p. 41.

[35] C. S. Reynolds, S. W. Wiseman, and M. J. O. Clarke, *J. Appl. Ecology*, 1984, **21**, 11.

[36] A. W. Fast, in *Destratification of Lakes and Reservoirs to Improve Water Quality*, ed. F. L. Burns and I. J. Powling, Australian Water Resources Council, Canberra, 1981, p. 515.

There is some controversy about the use of this technique in terms of its efficacy and, although it may be attractive in terms of costs, there are few lakes in the UK which are deep enough to warrant its use.

Bio-remediation

The recent increase in the understanding of biological processes in lakes has led to the development of 'ecotechnical methods' of manipulating the trophic status of lakes. The most widely used techniques of bio-manipulation involve artificial change in the abundance of predators to enhance grazing of phytoplankton by zooplankton.[37] The increase in grazing pressure reduces phytoplankton densities and results in improved water transparency.

There have been numerous experiments using this technique, mostly in shallow waters, in Europe and N. America. With the exception of the Norfolk Broads Restoration programme,[38] there have been few attempts to resolve eutrophication by bio-manipulation in UK lakes.

It is becoming increasingly clear that, in shallow lakes which have high rates of phosphorus sediment re-cycling, bio-manipulation is an essential element of successful restoration.

6 Management Framework

In the UK, management of eutrophication in lakes and rivers is driven by legislative requirements and by problems identified at specific sites. In England and Wales this work is managed and co-ordinated by the National Rivers Authority (NRA, now part of the Environment Agency). Although it is beyond the scope of this review to detail this work, it is relevant to mention the type of strategy needed and the influence of agriculture in this context.

The above description of eutrophication has illustrated the complex nature of the problem, particularly in relation to the influence of nutrients, the multiplicity of sources of phosphorus and the spectrum of its bio-availability. Clearly, the most effective long-term solution to many of our eutrophication problems will be to reduce the nutrient load to affected waters. However, it has also been shown that, because the concentrations of available phosphorus required to impose a control on primary production is very low (*e.g.* 5–10 μg l^{-1} total dissolved phosphorus), the reduction of nutrients from any one source alone is unlikely to be effective.

In order to provide a realistic approach to eutrophication control, the NRA has developed a framework for gathering the scientific evidence and for presenting the relevant information needed for consultation with all parties involved in resolving the problems. This framework includes the production of 'Action plans' for each water on a case by case basis. The principal reason for using this approach is because the NRA is attempting to persuade external

[37] J. Shapiro and D. I. Wright, *Freshwater Biol.*, 1984, **14**, 371.
[38] D. M. Harper, *Eutrophication of Freshwaters*, Chapman and Hall, London, 1992, ch. 8, p. 251.

organizations to incur expenditure to reduce eutrophication, and because there is likely to be a number of different solutions.

In the first stages of the development of an 'Action plan' all control options are considered. In the case of lakes, this process is aided by a PC-based 'expert system', PACGAP, which looks at the physical and chemical characteristics of the lake to determine the most likely option for control. Once further, more detailed information has been collected on the lake's nutrient inputs and other controlling factors, a more complex interactive model can be used (Phytoplankton RespOnse To Environmental CHange, PROTECH-2) to define the efficacy of proposed control options more accurately. This model is able to predict the development of phytoplankton species populations under different nutrient and stratification regimes.

The 'Action plans' will also include options for controlling nutrient inputs in the upstream catchments of affected stillwaters. These are likely to include reduction of nutrients from both point and diffuse sources and a range of different combinations aimed at reaching the target concentrations of nutrients required to achieve control in the receiving waters.

Once the options have been clearly defined it will be necessary to carry out a cost–benefit analysis of each option. This has two main objectives. First, the overall cost of the project will need to be assessed to determine whether or not it is financially viable and, second, to ensure that those who will be required to incur expenditure are fully aware of the commitment needed. The financial benefits to users of the waters for recreation, fisheries, navigation, *etc.*, are relatively easy to determine, but monetary valuation of the environmental benefits such as conservation and general amenity will be more difficult to define. As yet this aspect of the cost–benefit analysis has not been fully developed in the UK. Having determined a range of options and costs for eutrophication control in a particular catchment, consultation on the details of the 'Action plan' with all those involved is needed before any plan can be implemented.

7 Conclusions

The diffuse and point source inputs of nutrients from agricultural practices, causing eutrophication, are still ill-defined and the effects of nutrient regulation are difficult to predict. The options for control described, such as the construction of wetlands, reduction in fertilizer use, the introduction of buffer zones or the installation of treatment systems for dairy yard effluent, all need to be considered in cases where diffuse sources of nutrients are found to contribute biologically available phosphorus. Historically, varying degrees of success have been achieved following the implementation of a particular option or strategy for control. This highlights the need to continue studies not only of the practical aspects of control techniques, but also of the processes involved in the fate and behaviour of the nutrients.

Examination of the influence of agriculture on phosphorus input to freshwater requires careful study to determine the most appropriate control strategies. The influence will vary in relation to whether the phosphorus is dissolved or particulate. Also, we know that variation in the input of particulate and dissolved phosphorus varies with rainfall, fertilizer application rates, the history of the application of fertilizer, land use, soil type, and between surface and sub-surface water. The variation in phosphorus input with rainfall precludes short-term study of the sources, and careful planning of sampling strategies is needed to give an accurate account of these variations.

It is only following the collection and collation of nutrient data from all sources, including agriculture, the appreciation of control options, and the development and implementation of 'Action plans', that significant progress will be made with eutrophication control in the UK.

Impact of Agricultural Pesticides on Water Quality

KATHRYN R. EKE, ALAN D. BARNDEN
AND DAVID J. TESTER

1 Introduction

Pesticides are not new. The use of inorganic substances, such as copper, for controlling insects and diseases is mentioned in the Bible, but the first synthetic pesticides, the organochlorines, were not developed until the 1940s. Since then, hundreds of new pesticides have been produced to control a wide range of weeds, pests, and diseases. The problem of supplying good quality, cheap food for the world's increasing population seemed to have been solved. Unfortunately, problems were encountered with some pesticides, due to their build up in the aquatic environment and subsequent food chain, and therefore the pressure increased to regulate and monitor them more closely.

2 Legislation

Comprehensive legislation governs the use of pesticides in the UK. These controls are set out and implemented through Part III of the Food and Environment Protection Act 1985 (FEPA) and The Control of Pesticide Regulations 1986 (COPR).

The primary aims of FEPA are to:

(i) protect the health of human beings, creatures and plants;
(ii) safeguard the environment;
(iii) secure safe, efficient and humane methods of controlling pests; and
(iv) make information about pesticides available to the public.

These are achieved by prohibiting the advertising, sale, supply, storage or use of pesticides unless they have been granted approval by Government Ministers.

A pesticide is defined, under the Food and Environment Protection Act (1985), as 'any substance, preparation or organism prepared or used for destroying any pest'. Pesticides include herbicides, fungicides, insecticides, molluscicides, rodenticides, growth regulators, and masonry and timber preservatives.

As 'Guardians of the Water Environment', the National Rivers Authority (NRA) has statutory duties and powers to protect the aquatic environment from

K. R. Eke et al.

pollution. These duties are contained in the Water Resources Act (1991), which is the primary legislation for controlling and preventing water pollution. Under Section 85 of this Act it is an offence 'to cause or knowingly permit any poisonous, noxious or polluting matter or any solid waste to enter any Controlled Waters'. Controlled Waters are waters subject to the Water Resources Act (1991) and include all rivers, lakes, groundwaters, estuaries, and coastal waters.

Pesticides are potentially 'poisonous, noxious or polluting' substances, and therefore the NRA is responsible for controlling and preventing pesticide pollution of water. Furthermore, the NRA is responsible for ensuring water quality meets standards set in a number of EC Directives, some of which specify values for pesticides.

3 Pesticides and the Aquatic Environment

Currently there are about 450 pesticide active ingredients approved for use by the Ministry of Agriculture, Fisheries and Food (MAFF) and the Health and Safety Executive (HSE) in the UK, the majority of which are used in agriculture. Table 1 shows the 20 most used pesticides by weight in Great Britain for 1992 and 1994 and shows how the usage of pesticides varies considerably between years.[1]

Pesticides can enter the aquatic environment via a number of routes, including spillages, inappropriate disposal of dilute pesticides, and run-off into drains. Pollution from diffuse sources, such as spray drift into water courses and leaching from the soil, can also occur. In addition, there is evidence that some pesticides can be transported in the atmosphere over considerable distances.[2]

Pesticides vary widely in their chemical and physical characteristics and it is their solubility, mobility and rate of degradation which govern their potential to contaminate Controlled Waters. This, however, is not easy to predict under differing environmental conditions. Many modern pesticides are known to break down quickly in sunlight or in soil, but are more likely to persist if they reach groundwater because of reduced microbial activity, absence of light, and lower temperatures in the sub-surface zone.

4 Recognition of Pesticide Pollution

Pesticide pollution problems were first identified from the persistent organochlorine pesticides, such as DDT and dieldrin. At the time of its use, DDT was considered very safe and television footage shows DDT being sprayed onto women and children to highlight this. Unfortunately, DDT was not as safe as initially thought and the subsequent problem with it bioaccumulating through the food chain, resulting in the thinning of egg shells in birds of prey, is well known. Consequently, DDT was banned from use in the UK in the 1980s. Environmental concentrations have now generally declined to acceptable levels, but this has taken 10 years, due to its persistence in the environment. At a few heavily polluted sites, concentrations have yet to reach acceptable levels.

[1] MAFF/SOAFD Pesticide Usage Survey Group, *Arable Farm Crops in Great Britain 1994*, 1995.
[2] Water Research Council, *Atmospheric Sources of Pollution. Inputs of Trace Organics to Surface Waters*, R & D Report No. 20, Water Research Council, 1995.

44

Pesticide	1992	1994
Sulfuric acid	9994	12807
Isoproturon	2809	2339
Chlormequat	2413	2302
Mancozeb	1253	1054
Sulfur	679	1022
Chlorothalonil	910	760
Mecoprop	447	715
Glyphosate	261	567
Mecoprop P	539	510
Fenpropimorph	635	295
Carbendazim	289	287
Maneb	554	278
Fenpropidin	277	233
MCPA	220	232
Metamitron	234	229
Dimethoate	113	212
Trifluralin	327	206
Metaldehyde	78	185
Tri-allate	260	179
Carboxin	21	173

Table 1 The 20 most used pesticides by weight (tonnes) in Great Britain in 1992 and 1994

The first known incidence of pollution from approved herbicides was identified in 1972 in Essex, where tomato plants grown by commercial producers became malformed. The plants had been watered from public water supplies fed from a reservoir. The reservoir in turn abstracted water from a river supplemented by a water transfer scheme, from the River Cam in Cambridgeshire. Pollution from a factory manufacturing 2,3,6-TBA was identified as the cause and the problem was subsequently resolved by treating the effluent.

5 The Harmonized Monitoring Scheme

In 1974, the Harmonized Monitoring Programme was set up by the Department of the Environment (DoE). The objective was to provide a network of sites at the lower end of catchments, where water quality data could be collected and analysed in a nationally consistent manner, allowing the loads of materials carried through river catchments into estuaries to be estimated and long-term trends in river quality to be assessed. The complete list of substances to be monitored is diverse and specifies about 115 substances. The pesticides aldrin, dieldrin, γ-HCH, heptachlor, p,p'-DDT and p,p'-DDE are included. Figures 1 and 2 show the downward trend of γ-HCH and dieldrin over the past 20 years at the Harmonized Monitoring Sites. This confirms that reductions in environmental concentrations have been achieved, particularly over the past 10 years.

The concentrations of dieldrin show a marked decline throughout the 1980s following the ban on its use. The concentrations of γ-HCH have also declined during the same period, but the reduction is not as significant. This is probably due to the fact that γ-HCH still has a wide variety of uses in the UK in agriculture,

Figure 1 γ-HCH in rivers in Great Britain (% distribution of concentrations). (Data from Harmonized Monitoring Sites)

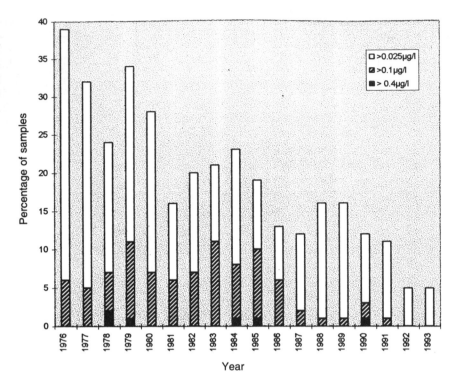

Figure 2 Dieldrin in rivers in Great Britain (% distribution of concentrations). (Data from Harmonized Monitoring Sites)

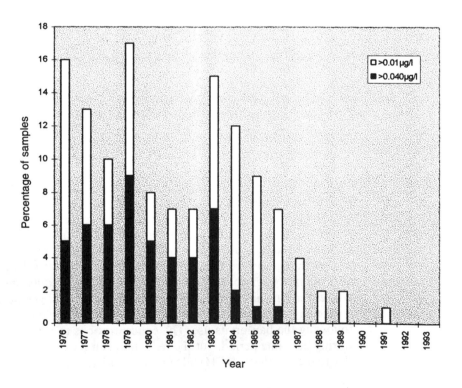

horticulture, forestry, and in human health products. It will be difficult to reduce concentrations further whilst γ-HCH is still approved for wide scale use.

6 Impact on the North Sea

In the 1980s, pollution of the North Sea was addressed by a series of ministerial conferences attended by countries bordering the North Sea. At the second conference, the UK Government agreed to halve the load of 36 priority hazardous substances entering the North Sea by 1995. Eighteen of these substances are pesticides. The substances were selected for priority control because of their toxicity, persistence in the environment, and potential to accumulate. The NRA is responsible, in England and Wales, for monitoring the inputs of these contaminants. The loads of these substances from industrial and wastewater discharges into estuaries and coastal waters are measured, and the load entering the sea at the tidal limit of rivers is assessed. Reduction targets have been achieved for most of those compounds arising from point source discharges by a mixture of tighter effluent discharge standards, new industrial technology, and the recession. It has proved much more difficult to reduce the loads arising from diffuse inputs, such as the normal use of pesticides without restrictions on use.

Figure 3 illustrates the loads of the pesticides simazine, trifluralin and γ-HCH entering the North Sea from England and Wales between 1990 and 1994. Because there were no baseline figures for pesticides in 1985, it is difficult to determine accurately whether loads to the North Sea have been halved, but it can be seen that significant load reductions have been achieved for simazine. This is probably due partly to the banning of its use on non-cropped land. Although the ban was not imposed until 1993, restrictions on the sale of simazine were imposed in 1992. The impending restrictions prompted users to switch to alternative products before this date.

The loads of trifluralin, however, increased in 1993 and again in 1994. It is not easy to explain why loads increased, since the MAFF usage figures indicate that trifluralin usage declined over the period. The loads for γ-HCH have declined since 1990, but are now starting to level out following the earlier decreases described above. Consents to discharge can be tightened to restrict discharges from point sources, but diffuse inputs from agriculture and from use in domestic situations, such as for the treatment of headlice, are much more difficult to control. It appears that the Government will need to consider new actions to reduce the loads of pesticides arising primarily from diffuse sources if the agreed reduction targets to are to be met.

7 Impact on Drinking Water Sources

In 1980 the Drinking Water Directive was introduced, which specified a maximum limit of 0.1 $\mu g\,l^{-1}$ for any pesticide in drinking water and 0.5 $\mu g\,l^{-1}$ for total pesticides. Monitoring was needed for a wide range of pesticides in water and this became the impetus for developing new analytical techniques capable of detecting pesticides at very low levels. Consequently, analytical techniques improved and more pesticides were detected in watercourses and water supplies,

Figure 3 Loads of simazine, trifluralin, and γ-HCH (tonnes/year) entering the North Sea from England and Wales during 1990–94

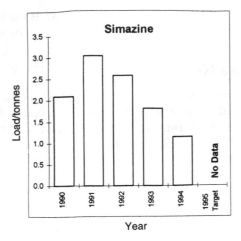

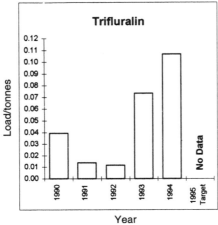

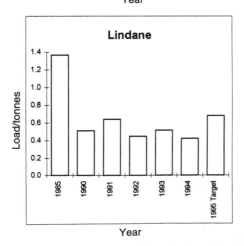

albeit at very low concentrations. The herbicides atrazine, simazine, isoproturon, and mecoprop were most frequently detected at levels above $0.1 \, \mu g \, l^{-1}$. The pesticides most frequently exceeding $0.1 \, \mu g \, l^{-1}$ in the NRA Anglian Region, a predominantly agricultural area, are illustrated in Figure 4.

Atrazine and simazine arose principally as a result of their use in amenity situations but, since their ban for non-agricultural purposes, concentrations are generally declining. However, atrazine and simazine still have some agricultural uses (atrazine on maize and simazine on a wide range of crops), so the risk of pollution still exists when these pesticides are applied in either groundwater or surface water drinking water supply catchments.

Problems have already been encountered in the South West of England, where the hectarage of maize is increasing. Atrazine applications increased and consequently atrazine was detected at concentrations above the Drinking Water Standard ($0.1 \, \mu g \, l^{-1}$) in a water supply river (R. Avon). A campaign was initiated, with the co-operation of the pesticide manufacturer, and farmers were targeted to remind them of ways to minimize water pollution. Some success resulted and concentrations were reduced the following year. Presently, satellite imagery is being used to identify maize fields, to enable campaigns to be targeted most effectively.

Isoproturon (IPU) is a widely used cereal herbicide for controlling blackgrass and is the most commonly applied agricultural pesticide in current use. The high tonnage used and the predominantly autumn application results in IPU contaminating surface and groundwaters. This is particularly problematic where the water is abstracted for drinking water. The solutions are either to install expensive treatment systems or close down the abstraction point, where this is possible, until the concentrations decline.

The problem of IPU contaminating water supply sources was highlighted on the Isle of Wight in March 1994. The public water supply intake on the River Eastern Yar had to be closed due to unacceptable levels of isoproturon and chlorotoluron. The NRA investigated and concluded that the pollution arose as a result of 'normal' agricultural spraying. The particularly wet autumn and winter resulted in most farmers concentrating all their IPU applications into three weeks in March. Consequently, the heavy rainfall following applications led to large flushes of IPU entering surface waters. This is illustrated in Figure 5.

IPU has recently been reviewed by the Advisory Committee on Pesticides, the independent scientific committee that advises ministers on pesticide approvals (ACP). The ACP concluded that IPU could continue to be used, subject to some restrictions on its use which are shown in Table 2. MAFF recommended that pesticides were not the only weapon against blackgrass and that cultural control methods, such as changing cultivation techniques and rotations could also help reduce blackgrass populations. Other recommendations withdrew the previous approval for using IPU before the crop had emerged (pre-emergence) and applications from the air. A product 'Stewardship' Campaign is being initiated by the IPU Task Force (a group of IPU manufacturers), including a leaflet on how to protect water quality. The NRA is keen to promote the leaflet and is willing to assist with the campaign, but is not convinced that these measures alone will be sufficient to reduce IPU concentrations in drinking water supplies to acceptable

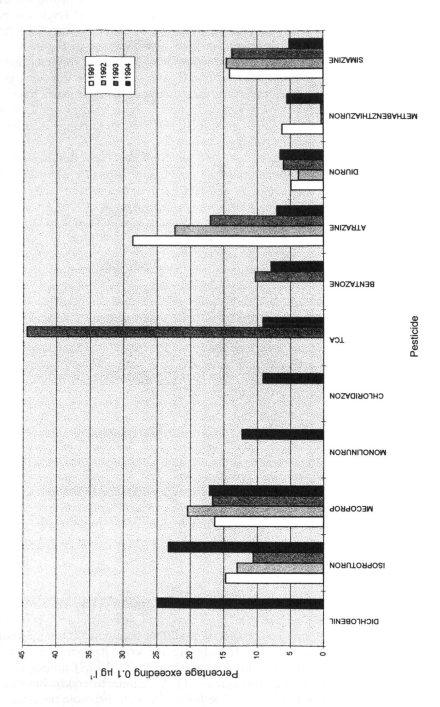

Figure 4 Pesticides most frequently exceeding 0.1 μg l^{-1} in Anglian Region, 1991–94

Table 2 Restrictions imposed on isoproturon following the ACP review

Emphasize cultural control measures for blackgrass
Investigate the role of set-aside for water protection and blackgrass control
Withdrawal of pre-emergence use
Impose a maximum dose rate of $2.5\,kg\,ha^{-1}$
Target rate of $1.5\,kg\,ha^{-1}$ when results of research are known
Not to use on cracked soils to avoid run-off through drains
Revocation of aerial use
Where a Water Protection Zone is created, use can be reconsidered within it

Figure 5 The concentration of isoproturon and associated rainfall at Burnt House on the River Eastern Yar, Isle of Wight, 1994

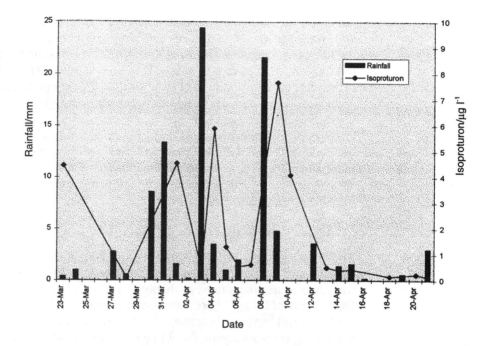

levels. The NRA was disappointed that MAFF failed to reduce the application rate to 1.5 kg/ha, which was considered one of the most effective ways of reducing IPU concentrations in surface waters. One recommendation put forward by MAFF which could help reduce IPU concentrations was that its use could be reconsidered within Water Protection Zones, where these have been created. Water Protection Zones can be set under Section 93 of the Water Resources Act (1991) and allow for the restriction or banning of a pesticide within these areas. Unfortunately, Water Protection Zones have not yet been set to control pesticides and their establishment will be difficult to justify until all other methods of control have been found to be ineffective.

Mecoprop is also found frequently in surface freshwaters. MAFF have recognized this and taken action. Mecoprop is a mixture of two different chemical forms, only one of which has herbicidal activity. Historically, products contained a mixture of both forms, resulting in the need for high doses to achieve the desired effect. However, it is now possible to separate the herbicidally active form 'mecoprop p' and MAFF's review concluded that only products containing the active form will be allowed. This halves the dose required and hopefully will reduce concentrations in watercourses. However, manufacturers have until December 1997 to submit data for new formulations and NRA data indicate that earlier action is needed. The NRA is trying to raise awareness amongst farmers and distributors, and is encouraging the use of 'mecoprop p' formulations whenever possible.

Bentazone has been monitored in the NRA Anglian Region since 1993 and the results show that bentazone is regularly present in surface and groundwaters. Currently there are no restrictions on its use, but bentazone is due to be reviewed under the 'Authorizations' Directive, the new European legislation for pesticide approvals, and the issue of water pollution will be raised.

The occurrence of pesticides at significant concentrations in groundwater is less frequent than in surface waters. However, extreme care must be maintained to prevent pesticide pollution of groundwater, for once polluted it is very difficult and expensive to remedy. In addition, evidence of high nitrate in groundwater sources in arable regions indicates that the area where pollution could occur is large. Pesticide exceedences of $0.1\,\mu\mathrm{g\,l^{-1}}$ in groundwaters are generally more significant in terms of drinking water than exceedences in surface waters, because most groundwater sources do not have the sophisticated treatment required to remove pesticides. The pesticides most frequently detected are the triazines, the urons, mecoprop and bentazone.

8 Impact on Estuaries and Coastal Waters

The agricultural pesticides detected in estuaries and coastal waters are primarily the same as those detected in surface freshwaters. Diuron, however, is found more frequently in marine waters, probably as a result of its use in antifouling paints. Exceedences of $0.1\,\mu\mathrm{g\,l^{-1}}$ are common for diuron, but because saline water is not used for drinking water supplies, this is not particularly significant. However, comparing pesticides with the $0.1\,\mu\mathrm{g\,l^{-1}}$ standard does give an indication of which pesticides are present in marine waters. Toxicity-based standards are required to determine the environmental risk to marine waters. These are discussed in Section 10. No environmental impacts from agricultural pesticides have become apparent in saline water as yet, but problems have been identified with the antifouling paints containing tributyltin (TBT), which has been shown to cause dog whelks to change sex. Restrictions on its use have subsequently been imposed.

9 Recent Pesticide Developments

In the 1990s, pesticide development progressed still further. The sulfonylurea herbicides were developed and have extremely low application rates. For example, the recommended rate for metsulfuron methyl is $6\,\mathrm{g\,ha^{-1}}$. This will undoubtedly reduce the likelihood of them being detected in water above $0.1\,\mu\mathrm{g\,l^{-1}}$, but does not necessarily preclude any environmental impact. Their high herbicidal activity means they may affect aquatic plants, even at very low concentrations. Monitoring for sulfonylureas is not yet undertaken by the NRA;

currently their low doses and predominantly spring applications results in monitoring being a lower priority than for some other pesticides. However, analytical techniques are currently being developed and special investigations will probably be undertaken to establish environmental concentrations in the near future.

Much of the NRA's current monitoring is targeted at herbicides because they are used in the largest tonnage and are frequently detected in water. Generally, however, insecticides and fungicides are more toxic to aquatic life than herbicides, and analytical techniques are being developed for those posing the greatest risk to the aquatic environment.

10 Environmental significance

To protect the aquatic environment and check that water is suitable for its recognized uses, such as abstraction for drinking, spray irrigation or fisheries, the NRA assesses water quality against Environmental Quality Standards (EQSs). An EQS is the concentration of a substance which must not be exceeded within the aquatic environment in order to protect it for its recognized uses. EQSs are specific to individual substances (including some pesticides) and are produced using the best available environmental and ecotoxicological information. Currently, EQSs only relate to surface water.

Statutory EQSs have been set in legislation by the European Commission (EC) and in the UK by the Department of Environment (DoE). Other non-statutory operational standards have been developed by the NRA to control discharges and assess water quality.

Some substances defined in List I of EC Directive 76/464/EEC have had statutory EQSs set in daughter Directives. These have been transcribed into UK legislation via the Surface Water (Dangerous Substances) (Classification) Regulations of 1989 and 1992. Any exceedence of these statutory EQSs downstream of relevant discharges are reported annually to the DoE and action is taken to prevent further exceedences. The only List I pesticide with current agricultural approval is γ-HCH and results indicate that a small number of sites fail to meet its EQS. Most are associated with polluted sites, but diffuse inputs are thought to be responsible for others.

Although statutory standards for List II substances have yet to be set in UK legislation, the Government Advice note Circular 7/89 sets out EQSs for a number of List II substances which are being applied as if they were statutory. Cyfluthrin and permethrin have agricultural approvals, but their detection in the environment is mainly associated with discharges from the moth-proofing industry.

In addition, the DoE proposed EQSs in a 1991 consultation document for those pesticides on the Red List (the UK's original priority hazardous substances list). Although non-statutory, the Government is committed to the reduction of Red List Substances discharging to the North Sea and the NRA uses the standards to assess the effects of these substances on the environment and to derive consents for point source discharges of these compounds. Failures for agricultural pesticides are rare.

The NRA is developing its own operational EQSs for priority pollutants

Table 3 Environmental quality standards and objectives for agricultural pesticides*

Pesticide	Freshwater			Saline water		
	Maximum $\mu g\,l^{-1}$	Annual average $\mu g\,l^{-1}$	95 percentile $\mu g\,l^{-1}$	Maximum $\mu g\,l^{-1}$	Annual average $\mu g\,l^{-1}$	95 percentile $\mu g\,l^{-1}$
HCH	—	0.1	—	—	0.02	—
Total (A + S)¶	10‡	2†	—	10‡	2†	—
Malathion	0.5‡	0.01†	—	0.5‡	0.02†	—
Trifluralin	20‡	0.1†	—	20‡	0.1†	—
Diazinon	0.1§	0.01§	—	0.15§	0.015†	—
Isoproturon	20§	2§	—	—	2§	—
Chlorotoluron	20§	2§	—	—	2§	—
Mecoprop	200§	20§	—	200§	20§	—
MCPA	20§	2§	—	20§	2§	—
Cyfluthrin	—	—	0.001†	—	—	0.001†
Permethrin	—	—	0.01†	—	—	0.01†

*Used by the NRA in 1993/94.
†Proposed by DoE.
‡Proposed by Water Research Centre.
§Proposed by NRA.
¶Atrazinie and Simazine.
Others are statutory.

through its national Research and Development Programme. Although non-statutory at present, these standards are used by the NRA to derive consent conditions. Table 3 shows some of the EQSs currently available for agricultural pesticides.

It is interesting to note that many of the agricultural pesticides exceed $0.1\,\mu g\,l^{-1}$, but very few have been recorded as failing these operational standards. In 1993 there were no failures in environmental waters for isoproturon, chlorotoluron, mecoprop or MCPA.

11 Monitoring

Pesticide monitoring has historically been targeted at pesticide problems known about for many years. This has now progressed and models are being used to predict the pesticides most likely to be detected in watercourses. Information on different catchments is available commercially from Farmstat Ltd in the form of the FARMSTAT report and some NRA Regions subscribed to this system in the early 1990s. A mathematical model predicts the likely concentrations of pesticides within specific catchments after taking into consideration rainfall, soil type, cropping, pesticide use, timing of application, and solubility.

Once pesticides were identified, monitoring was undertaken by the NRA, where possible, to confirm the usefulness of the model predictions. The most important prediction from the model was that the herbicide bentazone would reach surface waters. Subsequent analysis by the NRA confirmed the detection of bentazone at concentrations above $0.1\,\mu g\,l^{-1}$. Consequently, the NRA informed the water companies of the exceedences in rivers and advised them to monitor bentazone in drinking water supplies. It is unlikely that bentazone would have been included in NRA monitoring suites without the predictions, because it is used in fairly small tonnages and only applied in the spring. This indicates that there is no place for complacency with other pesticides that are not currently monitored for the same reasons (see Section 9).

Limitations with the system occur, however, because the model can only predict those pesticides which would have been expected in the previous year. The NRA is currently expanding this idea and producing a risk assessment tool which will predict the concentrations of pesticides in any catchment before the pesticides are applied. The system is known as POPPIE (Prediction of Pesticide Pollution In the Environment), and is an integrated computer system of relational databases, pesticide environmental fate and behaviour models, and Geographical Information Systems. POPPIE will integrate fully with the national pesticides database which contains concentrations of a large range of pesticides at sampling points in Controlled Waters, as well as data on sediment and biota concentrations. POPPIE will predict ground and surface water quality with respect to pesticides, help target monitoring programmes, and assess the leaching potential of new pesticides.

12 The Future

Improved targeting of pesticide monitoring will require the development of new analytical techniques and EQSs for priority pesticides to determine their environmental significance. To ensure confidence in the monitoring data, analytical techniques need to be able to detect the pesticide at 1/10th of the EQS, which can be as low as $1 \, \text{ng} \, \text{l}^{-1}$.

For the NRA to initiate effective control strategies to minimize the occurrence of pesticides in water, the source of the contamination needs to be known. If pesticides arise from true 'diffuse' use, *i.e.* the correct application of an approved pesticide, the NRA may need to request controls on the pesticide's use in order to meet the EQS. These may take the form of Water Protection Zones, as discussed in Section 7. Recent situations were appropriate controls are being investigated are for isoproturon on the Isle of Wight and atrazine on maize. Conversely, if pesticides arise from non-approved use, such as spillages or illegal disposal of pesticide washings, then improved training and education of users, coupled with increased policing of pesticide use, may help to prevent the problem.

Until recently, the NRA has not participated during the approval process in assessing the potential environmental impact of pesticides. However, the NRA does supply monitoring data to MAFF and HSE for pesticide reviews. These occur once a pesticide has been approved for use for a certain length of time, or when further information is needed on an approved pesticide. In supplying these data, the NRA comments on any areas of concern. This contributed to the 1993 ban on the use of atrazine and simazine on non-cropped land. In January 1995 the NRA's National Centre for Toxic and Persistent Substances (TAPS) was made advisor to the DoE, on the potential impact on the aquatic environment of

K. R. *Eke* et al.

the pesticide products being assessed by the ACP. This allows concerns over aquatic toxicity or fate and behaviour to be addressed at an early stage.

13 Conclusion

The full extent of the toxicity of pesticides to aquatic life is not known. Although chronic toxicity testing is required for new substances, little is known about the long-term effects of older pesticides. Also, very little is known about the toxicity and occurrence of the products formed when pesticides break down (metabolites) or the many non-pesticidal additives (co-formulants and adjuvants) used in pesticide formulations. However, the future is looking brighter. New modelling techniques, EQS development, and the involvement of the NRA in the pesticide registration process, coupled with the development of newer, less persistent pesticides with lower dose rates, all should help to reduce the risk of pesticide pollution.

Agricultural Nitrogen and Emissions to the Atmosphere

DAVID FOWLER, MARK A. SUTTON, UTTE SKIBA
AND KEN J. HARGREAVES

1 Introduction

The exchange of gaseous compounds of nitrogen between soil and the atmosphere has been a subject of scientific interest for more than a century. During the mid-19th century the focus of the debate was the source of nitrogen for agricultural crops, some scientists[1] believing that atmospheric inputs of gaseous nitrogen provided the plant nitrogen supply directly, while others[2] believed the soil was the major source.

The careful and long-term field measurements by Lawes and Gilbert[2] quantified the magnitude of the atmospheric inputs and showed that inputs from the atmosphere provided much less nitrogen than that removed from fields in non-leguminous crops. These measurements showed that the primary supply of nitrogen for fertilized crops was provided *via* the soil. While this appeared to settle the debate in favour of Lawes and Gilbert, the issue, like so many things, is more complex. Firstly, there are many ecosystems which rely in the long term almost exclusively on the atmosphere for their input of fixed nitrogen; ombrotrophic mires are a good example.[3] Such ecosystems, unless subjected to large inputs of nitrogen, lose very little by volatilization of gaseous oxidized or reduced nitrogen (N_2O, NO, or NH_3) to the atmosphere. The losses of gaseous nitrogen to the atmosphere occur primarily from ecosystems with a much larger supply, either resulting from fertilizer application, nitrogen fixation or atmospheric inputs in precipitation as NH_4^+ or NO_3^- or by the dry deposition of NO_2, NH_3, or HNO_3.[4]

The interest in gaseous losses of nitrogen from soil is now extensive and includes the well established community of soil scientists concerned with losses of fertilizer-applied nitrogen by nitrification and denitrification.[5] More recently, interest in ammonia losses from plants and soil has been stimulated by the very large emissions from intensive cattle production in the Netherlands and their

[1] J. Liebig, *Farmers Mag.*, 1847, **16**, 511.

[2] J. B. Lawes and J. H. Gilbert, *J. R. Agric. Soc.*, 1851, **12**, 1.

[3] R. S. Clymo, *Philos. Trans. R. Soc. London, Ser. B*, 1984, **303**, 605.

[4] U. Skiba, K. J. Hargreaves, D. Fowler and K. A. Smith, *Atmos. Environ., Part A*, 1992, **26**, 2477.

[5] K. A. Smith and J. R. M. Arah, *Losses of Nitrogen by Denitrification and Emissions of Nitrogen Oxides from Soils*, The Fertiliser Society, 1990, Proceedings No. 299.

Table 1 Deposition of NH_x to plate collectors (expressed as kg NH_x-N $ha^{-1}yr^{-1}$) recorded by Hall and Miller[11] at Rothamsted*

	Field (plot no.)							
	Lawn (1)		*Broadbalk* (7)		*Broadbalk* (12)		*Parkland* (4)	
Sampler height (m)	1.15	0.05	1.15	0.05	1.15	0.05	1.15	0.05
1908–1909	1.55	0.88	1.25	1.45	1.17	1.29	0.79	0.68
1909–1910	1.77	1.02	1.28	1.98	1.41	2.11	0.95	0.88

*Plate collectors were set at two heights above the ground, which may give an indication of air concentration gradients. The two Broadbalk plots received nitrogen fertilizers in the spring of each year; the lawn and parkland were unfertilized.

ecological consequences.[6] The effects include local ones, such as marked declines in heather and species diversity close to large NH_3 emissions,[7,8] and long-range effects associated with the regional transport and deposition of NH_4^+ aerosols across Europe. There is also strong evidence that the presence of NH_3 influences the rate of oxidation of SO_2 to SO_4^{2-} by heterogeneous processes in clouds, and that this influences the transport and deposition pathways of other major pollutants.

The interest in agricultural losses is not entirely new and important early measurements were available to alert the scientific community to the processes occurring. The work of Ville[9] and Schlösing[10] showed, in controlled conditions, that plants could absorb NH_3 from the atmosphere. The data available from field measurements designed to measure the atmosphere–surface exchange of NH_3 are less quantitative, but they show evidence of emission from fertilized grassland in the form of vertical gradients in the ambient concentrations of NH_3, with the largest concentrations close to the plant canopy.[11] For the unfertilized lawn the concentrations were smallest close to the surface, showing evidence of deposition (Table 1). Although of historical interest, these early measurements are not entirely satisfactory as the method was crude, based on the exposure of collector plates. In a vertical profile the measurements are confounded by the vertical gradient in wind velocity, leading to more efficient capture of NH_3 by the higher samples. Thus, with larger concentrations observed at the lower collector, it is reasonably certain that direction of the (emission) flux is correctly deduced even though the magnitude of the flux is uncertain. In the case of largest concentration at the higher level, the actual concentration profile may be very much smaller, or even of a different sign.

The early measurements therefore identified NH_3 volatilization, and were followed in the 1920s by studies which demonstrated NH_3 emission by vegetation.[12] The measurement of emission of N_2O and NO from agricultural sites has been

[6] G.J. Heij and J.W. Erisman, *Acid Rain Research: Do We Have Enough Answers?*, Elsevier, Amsterdam, 1995.

[7] J.G.M. Roelofs, *Experientia*, 1986, **42**, 372.

[8] G.W. Heil and W.H. Diemont, *Vegetatio*, 1983, **53**, 113.

[9] G. Ville, *C. R. Acad. Sci.*, 1850, **31**, 578.

[10] Th. Schlösing, *C. R. Acad. Sci.*, 1874, **78**, 1700.

[11] A.D. Hall and N.H.I. Miller, *J. Agric. Sci.*, 1911, **4**, 56.

[12] F.B. Power and V.K. Chesnut, *J. Am. Chem. Soc.*, 1925, **47**, 1751.

much more recent. In the case of N_2O[13] and NO,[14] the importance of soil NO emissions as a contributor to the global atmospheric budget of reactive nitrogen was recognized only within the last two decades.

The agricultural emissions of NH_3, N_2O and NO must be considered in context; the processes which lead to net loss from the soil and vegetation are natural and form a part of the land–atmosphere cycling of this vital nutrient. The current agricultural processes, however, create conditions in which the small natural background fluxes, in the range of a few $ng\,N\,m^{-2}\,s^{-1}$, are dwarfed by losses from fertilized land.

Early work[15] showed that denitrification processes led to significant losses of soil nitrogen. In this article we outline the major processes which lead to losses of oxidized and reduced gaseous nitrogen from agricultural soils and vegetation to the atmosphere. Where possible, the magnitude of the loss is scaled for the UK to show the magnitude of the fluxes relative to other well-known (pollutant) emissions.

2. Ammonia

In Europe, agriculture represents by far the largest source of ammonia emissions to the atmosphere. Although globally other natural and anthropogenic sources exist, such as the oceans, biomass burning, industrial combustion, humans, and wild animals,[16–18] in agricultural regions these sources are generally trivial relative to the volatilization of ammonia from livestock wastes and from fertilized arable land.[19–21] The main origin of the NH_3 emitted to the atmosphere is thus agricultural nitrogen, either from mineral nitrogen fertilizer or biological nitrogen fixation. The issue of NH_3 as a pollutant is coupled with the intensification of agriculture and increased nitrogen inputs over recent decades.

The main process of NH_3 emission centres on the decomposition of of livestock wastes containing either urea in animals or uric acid in poultry. In animals, most of the urea is contained in the urine, with the nitrogen in dung generally being present in organic forms and less liable to NH_3 emission. Hydrolysis of the urea by the enzyme urease, present on leaves and in the soil, produces NH_3 and CO_2. Since NH_3 is an alkali, its production provides a localized increase in pH, so that more of the total ammoniacal nitrogen is in the form of NH_3 rather than NH_4^+ ions. Ammonia emission takes place as a result of evaporation from liquid solution according to the Henry's law equilibrium between gas and liquid phase NH_3. Concentrations of gaseous NH_3 at the surface of volatilizing livestock waste may be of the order of 1–20 ppmv, much larger than the air concentration of NH_3 in the atmosphere, which are typically 0.01–10 ppb. Since these surface

[13] J.C. Ryden and J. Lund, *Soil Sci. Soc. Am. J.,* 1980, **44**, 505.

[14] I.E. Galbally and C.R. Roy, *Nature (London),* 1978, **275**, 734.

[15] R. Warrington, *J. R. Agric. Soc. England,* 1897, **8**, 577.

[16] E. Eriksson, *Tellus,* 1952, **4**, 215 and 296 (references).

[17] W.H. Schlesinger and A.E. Hartley, *Biogeochemistry,* 1992, **15**, 191.

[18] O.T. Denmead, *Aust. J. Soil Res.,* 1990, **28**, 887.

[19] E. Buijsman, H.F.M. Maas and W.A.H. Asman, *Atmos. Environ.,* 1987, **21**, 1009.

[20] W.A.H. Asman, *Ammonia Emission in Europe: Updated Emission and Emission Variations,* Report 228471008, RIVM, Bilthoven, The Netherlands, 1992.

[21] M.A. Sutton, C.J. Place, M. Eager, D. Fowler and R.I. Smith, *Atmos. Environ.,* 1995, **29**, 1393.

concentrations are so much larger than air concentrations, volatilization generally proceeds rapidly and without any influence by air concentrations. Efficient atmospheric mixing and dilution result in a rapid, near exponential, decline in air concentrations away from sources of emission, so that, even in the most polluted areas, mean annual concentrations of NH_3 rarely exceed 10–20 ppbv other than in the immediate vicinity (20–500 m) of major sources.

In the case of fertilized arable land, NH_3 emission may result from both direct fertilizer volatilization and from foliar emissions from vegetation as a consequence of the higher nitrogen status of the plants. The largest fertilizer emissions occur following the application of urea, which may lose 10–20% of the fertilizer nitrogen by NH_3 volatilization.[20,22,23] The process of emission is similar to that for livestock waste, with the large emissions resulting from the pH increase occurring during hydrolysis. Other more neutral fertilizers, such as ammonium nitrate, provide much smaller emissions (*e.g.* <1% applied N for ammonium nitrate).

Ammonia emissions from fertilized plant foliage may occur throughout the life cycle of a crop and are closely tied to nitrogen metabolism and plant phenology.[24] Peaks in emission may occur, because the plants have excess nitrogen in the weeks following fertilization, and also after anthesis as protein breakdown in leaves occurs and nitrogen is remobilized for transfer to the grain.[25,26] Again the process is one of evaporation of NH_3 from solution. Ammonia and ammonium are important components of plant nitrogen metabolism, and exist in the apoplastic (intercellular) solution of leaves. Hence factors affecting either apoplastic pH or ammonium concentrations will affect the potential for emissions from crop leaves, which take place largely through stomata. In contrast to volatilization from livestock wastes, the equilibrium concentrations of NH_3 at the emitting surface are much smaller, typically in the range 0.5–40 ppb, so that it is much more typical for fluxes with crops to be bi-directional, with periods of both emission and deposition. Other sources of NH_3 emission from fertilized plants include decomposition of fallen decaying leaves (such as in oilseed rape, or where grass is cut and left in the field).

The foregoing discussion identifies the range of physiological as well as solution equilibria and deposition processes which combine to regulate the net exchange of NH_3 between soils, vegetation, and the atmosphere. The processes are summarized diagrammatically in Figure 1. The extent to which current understanding of the processes may be used to calculate fluxes of NH_3 between canopies of vegetation and the atmosphere is described by Sutton *et al.*[27] The approach requires a knowledge of apoplast NH_4^+ concentration, stomatal resistance, and the atmospheric resistances r_a and r_b.[28] The detailed procedure

[22] S. G. Sommer and C. Jensen, *Fert. Res.*, 1994, **37**, 85.
[23] ECETOC, *Ammonia Emissions for 16 Countries in Europe*, European Centre for Ecotoxicology and Toxicology of Chemicals, Brussels, 1994.
[24] M. A. Sutton, D. Fowler, J. B. Moncrieff and R. L. Storeton-West, *Q. J. R. Meteor. Soc.*, 1993, **119**, 1047.
[25] L. A. Harper, R. R. Sharpe, G. W. Langdale and J. E. Giddens, *Agron. J.*, 1987, **79**, 965.
[26] J. K. Schjørring, A. Kyllingsbaek, J. V. Mortensen and S. Byskov-Nielsen, *Plant, Cell Environ.*, 1993, **16**, 161.
[27] M. A. Sutton, J. B. Moncrieff and D. Fowler, *Environ. Pollut.*, 1992, **75**, 15.
[28] J. Monteith and M. H. Unsworth, *Principles of Environmental Physics*, Edward Arnold, London, 1990.

Figure 1 Proposed NH$_3$
cycles in the soil–vegetation–
atmosphere system. (Taken
from Sutton *et al.*[27]).

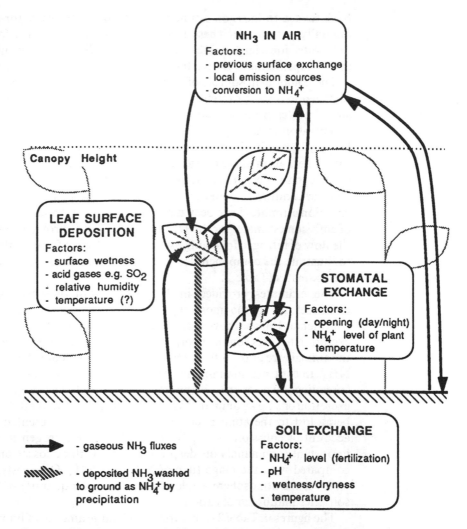

lies outside the scope of this article but it provides a mechanistic basis for
predicting the net crop–atmosphere exchange of NH$_3$ in field conditions. The
current limitation on the application of the method to calculate landscape-averaged
fluxes over seasonal timescales is the lack of appropriate input data. In practice,
scaling such fluxes over the agricultural landscape will require simplification of
the mechanism to obtain fluxes from readily obtained crop and atmospheric
variables and will probably be achieved using Geographical Information System
(GIS) methodology.

Quantities of Ammonia Emitted from Agricultural Sources

Providing quantitative estimates of NH$_3$ emissions is necessarily rather uncertain,
because of the wide range of variability in the sources as well as in factors affecting
the emission rate. In the case of livestock emissions, the obvious scaling
parameter is the number of animals in a particular area and emissions estimates

have tended to focus on estimating average 'emissions factors' of NH_3 emitted annually per animal. These have generally been provided from experimental data, often for a large number of animals and dividing by the number of animals in the experiment to derive the emission per animal.

The component sources of NH_3 emission from animals relates to the flow of nitrogen following excretion. Emissions may occur from animal houses, from stored wastes, from application to land or directly from fields for grazing animals. A central observation is that NH_3 emissions per animal are estimated to be much larger where animals are housed (housing, storage, and land spreading emissions) than where animals are outside grazing. In the case of cattle, which are the major source, this is despite the fact that animals graze during summer months, when warmer conditions would tend to promote emissions. A large number of NH_3 emission estimates have been made recently,[20,21,23] and have provided a review of emission estimates for the UK.[23] Sutton *et al.*[21] provided a simple scheme for the flow of nitrogen following excretion of nitrogen from cattle which may serve to illustrate the component losses of NH_3 emission as well as the uncertainty in the estimates (Table 2).

The estimates provided in Table 2 are not definitive and much debate continues as to the most representative estimates, both for the UK and continental Europe; however, these clearly show the flow of nitrogen following excretion to emission or incorporation. Overall, for the summer months where the animals are grazing, 9% of the N excreted is estimated here to be volatilized as NH_3. In contrast, summing up the emissions due to housing, storage, and land spreading of wastes provides an estimated loss of 14 kg N animal^{-1} yr^{-1} out of an excretion of 32.5 kg animal^{-1} yr^{-1}, which represents over 40% of the excreted N being lost to the atmosphere as NH_3. The figures presented here do not take account of emissions as N, N_2O or NO in the nitrogen flow, which may be important for animals on deep litter, though these losses are generally small compared with the magnitude of N lost as NH_3. Obviously, these additional losses to the atmosphere would result in a smaller quantity of N being left in the soil after manure application.

The figures in Table 2 summarize a very large amount of information, and such data vary with different animal classes as well as animals on different diets.[23] Thus cattle on intensive high-nitrogen diets, such as in the Netherlands or in S. Sweden, are expected to emit more NH_3 than cattle provided with lower nitrogen diets. Larger emissions are thus expected from dairy animals than from beef, although other differences also exist, such as between animals on slurry or solid waste systems (farmyard manure plus straw). Given the link between NH_3 emissions, solubility and temperature, the magnitude of the emissions factors will also depend on climate. Factors including wetness, temperature, seasonality and turbulence, as well as soil factors such as pH, cation exchange capacity, and drainage, will all influence emissions, making extrapolation to regional or continental scales increasingly less certain. Most of the experimental information defining the emissions derives from N.W. European experience,[29] and hence estimates are optimized for this region.

[29] S. C. Jarvis and B. F. Pain, *Proc. Fert. Soc.*, 1990, **298**, 35.

Table 2 Ammonia emission factors from cattle in the UK

Term	N available for emission and uncertainty (kg N animal^{-1} yr^{-1})	Best loss esimate and uncertainty (% of available N)	Equivalent component NH$_3$ emission factor (kg N animal^{-1} yr^{-1})
Annual N excretion from cattle	65 (40–90)	—	—
Field emissions (half of year)	32.5 (20–45)	9*	3 (2–4)
Housing emissions (half of year)	32.5 (20–45)	15*	5 (3–7)
Waste put into store	27.5 (17–38)†	10 (5–15)‡	—
Waste applied to land	25 (16–32)*,†	25 (15–30)	—
Combined emissions from storage and land spreading	—	33 (10–41)	9 (3.3–16)§
Total ammonia emission	—	—	17 (8.3–27)
Applied N not volatilized	18 (14–22)†	—	—

*Values inferred from component emission factors and given as indicators only.
†Uncertainty estimates assume that maximum emissions are associated with large N availability.
‡Conservative estimates based on very limited experimental information. Assumes half waste is slurry with 5% loss and half is farmyard material with 15% loss.
§Estimated from combined % losses.

Estimates of Ammonia Emissions in the UK

The comparative magnitude of the different sources of total NH$_3$ emissions may be illustrated by the estimates of emissions for the UK of Sutton *et al.*,[21] shown in Table 3. Consistent with other estimates for this country,[23,29,30] the estimates here show that emissions from cattle dominate the estimated emission, here estimated at a contribution of over 50% of the total. In comparison, sheep, pig, poultry and fertilizer/crop emissions are estimated to contribute similar amounts of around 8–9% each towards the total. Overall, including minor agro-industrial sources, agriculture is estimated to contribute around 90% of the total, with the remainder from miscellaneous non-agricultural sources. Greater attention has been given to these other sources in recent years.[31,32] The estimates for each of these small sources are highly uncertain; however, with the information available, it is difficult for these to account for anything but a small fraction of the total. Of the 'non-agricultural' sources, the most likely term which may be underestimated, and that will increase in the future, is emissions from sewage works and land spreading of sewage sludge. More information is required on this term.

[30] M. Kruse, H. M. ApSimon and J. N. B. Bell, *Environ. Pollut.*, 1989, **56**, 237.
[31] H. S. Eggleston, in *Ammonia Emissions in Europe: Emission Coefficients and Abatement Costs*, Proceedings of a Workshop 4–6 February 1991, ed. G. Klaassen, IIASA, Laxenburg, Austria, 1992, p. 95.
[32] D. S. Lee and G. J. Dollard, *Environ. Pollut.*, 1994, **86**, 267.

Table 3 Estimates and uncertainty of ammonia emission in the UK

Source	Population (thousands)	Average emission factor (kg N individual^{-1} yr^{-1}or % where noted)	Contribution to UK emission (Gg NH$_3$ yr^{-1})	Percent contribution (%)
Agricultural sources				
Cattle	11 872[a]	17 (8.3–27)	245 (119–389)	54.4
Sheep	40 942[a]	1.1 (0.4–1.5)	41 (15–56)[b]	9.1
Pigs	7 980[a]	4.3 (3.1–5.7)	42 (30–55)	9.3
Poultry (fowls)	130 808[a]	0.22 (0.15–0.3)	35 (24–48)	7.8
Turkeys and geese	8 110[a,d]	0.7 (0.48–0.95)[c]	7 (5–9)	1.6
Ducks	1 624[a]	0.1 (0.06–0.13)[c]	0.2 (0.1–0.3)	0.0
Fertilizer application and crops	—	2.4 (1.5–5)	34 (21–69)[f]	7.6
Agro-industrial sources	—	—	1.6 (0.8–3.3)	0.4
Non-agricultural sources				
Human sweat and breath	57 000	0.04 (0.01–0.11)	3 (0.7–7)	0.7
Horses	550[e]	10 (5–20)	7 (3–13)	1.6
Dogs	7 800[g]	0.81 (0.29–1.1)	7.6 (2.7–10.4)	1.7
Cats	7 100[g]	0.13 (0.05–0.18)	1.1 (0.4–1.6)	0.2
Land spreading of sewage sludge	—	—	11 (3–20)	2.5
Sewage works	—	—	1.5 (0.8–3)	0.3
Landfill, transport and coal combustion, waste incineration	—	—	9–11 (5–20?)	2.2
Biomass burning	—	—	2 (0.3–8)	0.4
Natural soils	—	—	0[h]	0
Wild animals (deer + sea birds)	—	—	1 (0.3–2)	0.2
Sum agricultural	—	—	406 (215–630)	90.2
Sum non-agricultural	—	—	44 (16.2–85)	9.8
Total			450 (231–715)	100

[a]Based on livestock statistics for 1988 (MAFF, 1990).[33]
[b]Corrected for six-month lifespan of lambs, which account for 50% of June census sheep numbers.
[c]Based on Aman, 1992,[20] and uncertainty in poultry emission factor.
[d]Estimated from available data.
[e]Asman, 1992.[20]
[f]For arable land, plus urea emissions from pasture. Based on total UK fertilizer consumption (Asman, 1992)[20] and 0.68 agricultural land area as arable and ungrazed grass (MAFF, 1990).[33]
[g]Lee and Dollard, 1994.[32]
[h]Any minor temporary emissions treated in definition of net dry deposition.

Nevertheless, in the same way that cattle dominate the total emission, these also dominate the uncertainties, so that most attention is obviously required here.

The estimates in Table 3 represent one inventory; however, a further estimate of uncertainties may be found by comparing other recent ongoing inventories for the UK. Pain *et al.*[34] have very recently estimated UK emissions from livestock production systems (including tillage crops) at 240 Gg NH_3 yr^{-1}, while ApSimon *et al.* (personal communication) estimate a figure of around 280 Gg NH_3 yr^{-1}. Including the non-agricultural sources at around 40 Gg NH_3 yr^{-1}, the 'official' estimate recently adopted by the Department of the Environment is 320 Gg NH_3 yr^{-1}. If these estimates are contrasted with the other recent estimate of UK emissions by ECETOC[23] of 594 Gg NH_3 yr^{-1}, it becomes evident that there is still much debate over these estimates.

Trends in Ammonia Emissions Over Recent Decades

Despite the uncertainties in the absolute magnitude of NH_3 emissions, it is still possible to use the available information to address the question of how emissions have changed over the past century. Using the inventory approach of scaling emissions by animal numbers, Asman *et al.*[35] provided emissions inventories of NH_3 for European countries for the years 1870, 1920, 1950, and 1980. These inventories were used as inputs to a model of the atmospheric transport and deposition of NH_3 over Europe. Over the same period, wet deposition of ammonium was measured at Rothamsted,[36,37] which provides a record with which to compare the model estimates.[38] These results are presented in Figure 2, and show both the modelled wet deposition (for this part of the UK) and the measurements at this site increasing, particularly since 1950. The measurements actually increase much more than the model, which may be due to local effects or because of simplifications in the modelling. Asman *et al.*[35] scaled their emissions solely on the basis of changes in animal numbers and fertilizer use. Hence, agricultural intensification since the 1950s will be expected to have provided increased emissions per animal, which may explain the increase of the measurements over the model.

Spatial Distribution of Ammonia Emissions

Assessing the spatial distribution of NH_3 emissions is of particular interest because of the link with ecological impacts of nitrogen deposition. Using statistical atmospheric transport models, such emission maps may be used to

[33] MAFF, *Agricultural Statistics, United Kingdom, 1988*, Ministry of Agriculture, Fisheries and Food, London, 1990.

[34] B. F. Pain, T. van der Weerden, S. C. Jarvis, B. J. Chambers, K. A. Smith, T. G. M. Demmers and V. R. Phillips, *Ammonia Emission Inventory for Agriculture in the UK. Final Report*, CSA2141/OC9117, Institute of Grassland and Environmental Research, North Wyke, Okehampton, UK, 1995.

[35] W. A. H. Asman, B. Drukker and A. J. Janssen, *Atmos. Environ.*, 1988, **22**, 725.

[36] P. Brimblecombe and J. Pitman, *Tellus*, 1980, **32**, 261.

[37] K. W. T. Goulding, P. R. Poulton, V. H. Thomas and R. J. B. Williams, *Water Air Soil Pollut.*, 1986, **29**, 27.

[38] M. A. Sutton, C. E. R. Pitcairn and D. Fowler, *Adv. Ecol. Res.*, 1993, **24**, 302.

Figure 2 Comparison of measured wet deposition of ammonium at Rothamsted, England with model estimates by Asman *et al.*[34] for regions which assume changes in emissions are only due to differences in animal numbers. (Taken from Sutton *et al.*[38]).

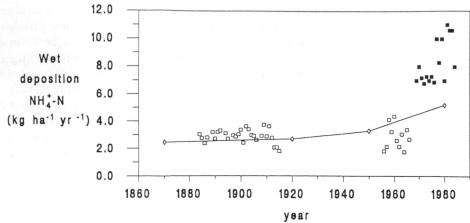

model NH_3 concentrations in the air as well as deposition to ecosystems. In principle, direct measurements of air concentrations are to be preferred to those inferred from a combination of model emissions and atmospheric transport modelling; however, because NH_3 emissions and air concentrations are expected to be very variable, measurements on their own would require an exceedingly large number of monitoring locations to reproduce the spatial variability of the concentration field. Both methods are therefore useful, and estimation of reliable NH_3 emissions and spatial distribution becomes an important component in quantifying nitrogen deposition to ecosystems.

The spatial resolution is the key constraint in such mapping exercises. Ideally, the resolution of the model should match the variability that occurs in practice. For slowly formed secondary pollutants, such as ammonium aerosol, this may not be too difficult. However, since NH_3 is emitted by ground level sources and deposits rapidly to sensitive ecosystems, the variability exceeds that which can be mapped on regional scales. Substantial gradients are expected over distance scales of 10–500 m, which would be particularly relevant for point sources, as well as manured fields. In mapping NH_3 emissions over regional scales, the main scalars are data on animal numbers and crop areas. Such information is generally collected at a parish (in the UK) or municipality level by agricultural ministries, and farm-by-farm information is not generally available. Mapping emissions on a regular grid is, therefore, subject to input data on variable spatial scales in relation to parish sizes. In the UK, parish sizes are typically 2–10 km × 2–10 km in size, so that the finest practical resolution is in the region of 5 km × 5 km. Conveniently, the areas with larger parish sizes are generally extensive upland areas where the spatial variability in emissions is less of a problem. Procedures for disaggregating NH_3 emissions and for mapping at 5 km grid resolution have been described by Eager[39] and include mapping the parish emissions onto 1 km grid land-use maps before re-aggregating up to a 5 km × 5 km scale. An example of

[39] M. Eager, MSc Thesis, University of Edinburgh, UK, 1992.

the derived 5 km maps of NH_3 emission for the UK is presented in Figure 3,[21] and includes emissions from the agricultural and non-agricultural sources shown in Table 3. The small contribution by non-agricultural sources was for most categories scaled by human population.

Figure 3 shows the substantial spatial variability in NH_3 emissions, particularly in source regions. It is clear that maps at a lower resolution, which are often made (*e.g.* 20 km grids for the UK,[40] 150 km grids for Europe[41]), will artificially smooth out emissions. The influence on the assessment of nitrogen deposition effects, such as using critical loads comparisons[42] of those scaling issues, will be substantial as dry deposition of NH_3 may be much more patchy than indicated previously. Nevertheless, the 5 km resolution shown here is still uncertain and substantial sub-grid variability will still occur. The uncertainty in locating the sources at finer scales than 5 km × 5 km, is likely to prevent national mapping at a finer scale. However, there is still an important role for local case studies examining spatial variability at scales matching that which occurs in practice (e.g. 100 m grid resolution), which may provide the basis for a statistical estimate of deposition and effects at a finer grid scale than 5 km × 5 km. Such studies are also needed to identify the magnitude of error that may be expected with regional estimates at limited resolution.

The Atmospheric Budget of Fixed Nitrogen Over the UK

The national (UK) emissions of NH_3 contribute to the air chemistry of the lower atmosphere, NH_4^+ being the dominant non-marine cation in precipitation over much of the country.[43] The dominance of NH_4^+ in precipitation reflects the anthropogenic aerosol composition over the country in which NH_4^+, SO_4^{2-}, NO_3^-, and H^+ represent the major components. Gaseous NH_3 is readily incorporated into acidic cloud or rain droplets, where it is important in maintaining a high pH of cloud water and, through this process, influences the pathway for oxidation of SO_2 to SO_4^{2-}. Gaseous NH_3 is also readily incorporated into acidic aerosols and the reactive nature of NH_3 with the ground and with acidic aerosols, cloud or rain, leads to a short atmospheric lifetime as a gas of a few hours.

As a consequence of the short atmospheric lifetime of NH_3 gas, the majority of the 'export' of reduced nitrogen from the atmosphere over the UK is as aerosol from the coast, and predominantly only from the east coast.

The measurement data for reduced atmospheric nitrogen in the UK are of variable quality, largely because monitoring ambient NH_3 is subject to very large variations.[43] However, the wet deposition monitoring data provide a good

---

[40] INDITE, *Impacts of Nitrogen Deposition in Terrestrial Ecosystems in the United Kingdom*, Department of Environment, London, 1994.

[41] H. Sandnes and H. Styve, *Calculated Budgets for Airborne Acidifying Components in Europe, 1985, 1987, 1988, 1989, 1990 and 1992*, EMEP Report 1/92, Norwegian Meteorological Institute, Oslo, Norway, 1992.

[42] CLAG, *Critical Loads of Acidity in the United Kingdom*, Summary Report of the Critical Loads Advisory Group, ITE, Edinburgh, 1994.

[43] RGAR, *Acid Deposition in the United Kingdom 1986–1988*, Third Report of the United Kingdom Review Group on Acid Rain, Department of Environment, London, 1990.

Figure 3 Estimated annual ammonia emissions for the UK including both agricultural and non-agricultural sources. (Taken from Sutton *et al.*[21]).

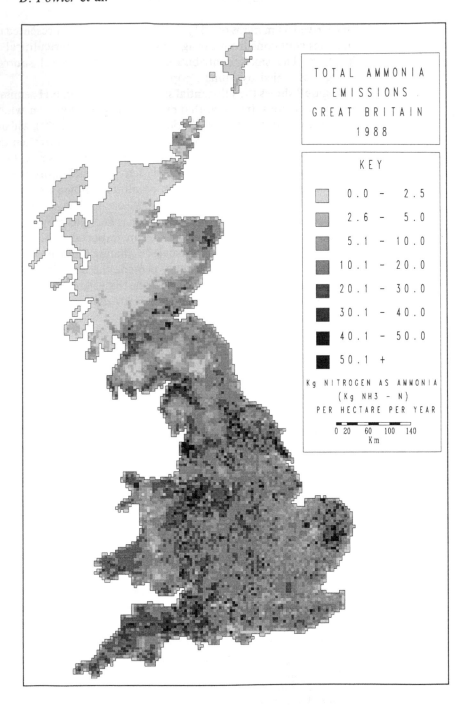

TOTAL AMMONIA
EMISSIONS.
GREAT BRITAIN
1988

KEY

0.0 - 2.5
2.6 - 5.0
5.1 - 10.0
10.1 - 20.0
20.1 - 30.0
30.1 - 40.0
40.1 - 50.0
50.1 +

Kg NITROGEN AS AMMONIA
(Kg NH3 - N)
PER HECTARE PER YEAR

0 20 60 100 140
Km

estimate of the annual wet deposition throughout the country at 130 kt NH_3-N.[40] The dry deposition, while uncertain due to NH_3 concentration limitations in the measurement methods, has been estimated at 100 kt NH_3-N. The total annual emission of NH_3 in the UK amounts to about 350 kt NH_3-N, so that the deposition of reduced nitrogen represents approximately 230/350 kt at 66% of emissions annually.[40] The atmospheric budget is only very approximate and does not include the input of NH_4^+ aerosol in southerly and easterly winds; however, the export flux implied by the budget of about 80 kt NH_3-N may be shown to be consistent with measured aerosol concentrations and the coastal flux. The measured aerosol NH_4^+ concentration at High Muffles on the east coast averaged 1.0 μg NH_4-N m^{-3} for 1988.[43] Taking a coast length of 500 km as representative of the coast for the NH_3 emitting area of the country, 1 km as the boundary layer height and 5 m s^{-1} as wind speed, the annual export would be about 80 kt NH_3-N. Other locations in East Anglia provide larger aerosol NH_4^+ concentration, but this simple budget shows that the probable export is of the same order as the discrepancy in the UK NH_x budget.

The individual terms in the atmospheric budget of reduced nitrogen are summarized in Figure 4, along with the equivalent components of the budget for oxidized nitrogen. The budget, while uncertain in several important terms, shows that the deposition of reduced nitrogen in the UK is primarily of UK origin, so that any problems due to NH_3 in the UK are largely a domestic problem, and that a relatively small fraction (33%) of the deposition is exported. The contrast with oxidized nitrogen is striking, in that the majority (74%) of the nitrogen oxides, which are largely from industrial and vehicle sources, are exported. The international 'exchange' of oxidized nitrogen is, therefore, much more important and control of this problem in the UK could not be achieved so readily by domestic regulation alone. It is also of interest to contrast the total nitrogen deposited in the UK with fertilizer use. The current fertilizer usage in the UK is between 1200 and 1400 kt N, and is of the same order [1200 kt $(NO_x + NH_3)$-N] as the combined emissions of oxidized and reduced nitrogen. The total deposition of nitrogen in the UK at 450 kt represents 30% of the fertilizer application. These comparisons are, of course, simply a means of communicating the relative magnitudes of the atmospheric and fertilizer nitrogen quantities. In practice, the reduced nitrogen emissions represent a fraction of the fertilizer applied that has been utilized by vegetation, consumed and excreted by the farm animals and emitted to the atmosphere. The recent trends towards larger, intensive animal production units, and also in the specialization of specific regions of the country for this form of agriculture, all tend to increase the fraction of applied nitrogen emitted to the atmosphere, as a smaller fraction of the emission is captured in the immediate vicinity of the emission source.

The concern over ecological consequences of the atmospheric input of oxidized and reduced nitrogen centres on soil acidification by the oxidized nitrogen directly, and by the reduced nitrogen following its transformation in the soil into NO_3^- or due to its uptake by vegetation (Figure 5).[44] The other major area of concern centres on the eutrophication of semi-natural

[44] P. A. van Breeman, E. J. Burrough, H. F. Velthorst, T. van Dobben, T. B. De Wit, X. X. Ridder and H. F. R. Reijnders, *Nature (London)*, 1982, **299**, 548.

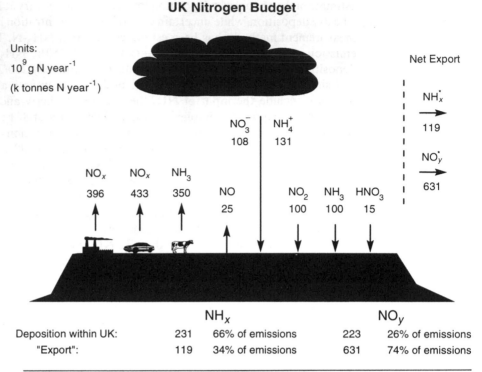

Figure 4 UK nitrogen budget, 1986–1988.

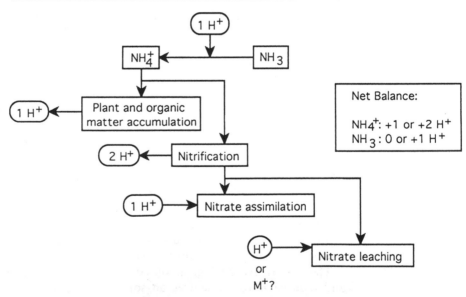

Figure 5 Effects of atmospheric NH_3 or NH_4^+ input on soil proton balances. The production of H^+ depends on the form of NH_x input and the fate of the NH_4^+ ion. M^+ represents a metal ion that might be leached from the soil. (Taken from Sutton et al.[38]).

plant communities by atmospheric nitrogen inputs.[8] In the case of agricultural sources, the NH_3 volatilized from individual fields or stock-rearing units is readily deposited onto shelter belt woodland or nearby unfertilized land. The nitrogen supply then presents an opportunity for plant species which are adapted

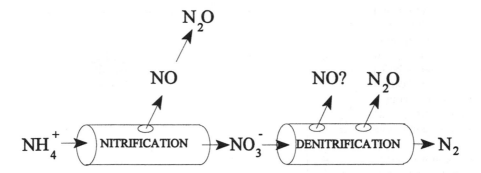

re 6 The production and
ission of NO and N₂O
during nitrification and
denitrification.

to respond to the raised nutrient supply to out-compete slow-growing species which have adapted to survive in nutrient-poor conditions. There is evidence that changes in species composition have been taking place on a regional scale throughout the UK[45] and that larger changes have been observed elsewhere in Europe.[8,46]

3 Emissions of NO and N₂O from Agricultural Soils

Nitrification and Denitrification

In soil, microbial nitrification and denitrification are the predominant sources of NO and N_2O and the emission fluxes may be regarded as leakage during the transformation processes shown in Figure 6. Nitrifiers can produce NO and N_2O during the oxidation of NH_4^+ to NO_3^-. Both gases are by-products of the nitrification pathway and the typical yield of NO in well-aerated soil is 1–4% of the NH_4^+ oxidized and for N_2O is less than 1%.[47,48]

Nitric oxide and N_2O are direct intermediates in the denitrification pathway, the reduction of NO_3^- to N_2. Reduction to N_2 is often incomplete, so that both N_2O and N_2 are equally important end products of denitrification, the ratio of N_2O/N_2 production being determined by soil physical properties. For example, N_2O is the main end-product in acid soils, whereas low redox potentials and high organic matter content favour the further reduction to N_2.[49] Nitric oxide production under anaerobic conditions (optimal denitrification conditions) was shown to be 1–2 orders of magnitude greater than under aerobic conditions (optimal for nitrification to occur).[50] The net release of NO from a soil, however, is greatly influenced by the diffusion properties and the rate of NO consumption in the soil. When a soil or microsite within the soil is sufficiently reduced for denitrification to occur, it is unlikely that the gaseous products escape to the atmosphere very easily and the chances of NO being consumed by the denitrifying community is very high. The difference between production and

[45] C. E. R. Pitcairn, D. Fowler and J. Grace, *Environ. Pollut.*, 1995, **88**, 193.
[46] U. Falkengren-Grerup, *Oecologia*, 1986, **70**, 339.
[47] E. J. Williams, A. Guenther and F. C. Fehsenfeld, *J. Geophys. Res.*, 1992, **97**, 7511.
[48] E. J. Williams, G. L. Hutchinson and F. C. Fehsenfeld, *Global Biogeochem. Cycles*, 1992, **6**, 351.
[49] W. J. Payne, in *Denitrification, Nitrification and Atmospheric Nitrous Oxide*, ed. C. C. Delwiche, Wiley, New York, 1981, p. 85.
[50] A. Remde, F. Slemr and R. Conrad, *FEMS Microbiol. Ecol.*, 1989, **62**, 221.

Figure 7 The production and emission of NO during denitrification in agricultural soil treated with NO_3 fertilizer (KNO_3) and the nitrification inhibitor Dyciandiamide (10%) under aerobic (air) and anerobic conditions (N_2). Fluxes are means from three soil columns, error bars represent standard deviations from the mean. V = vertical flow through the column; H = Horizontal flow over the soil surface.

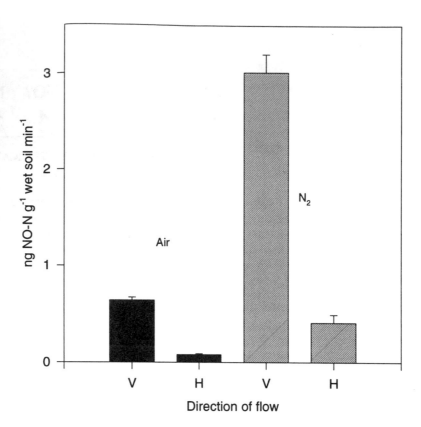

emission of NO during denitrification was demonstrated by the following experiment. Agricultural soil (200 g) was mixed with KNO_3 (0.04 mg N g^{-1} soil) and the nitrification inhibitor Dyciandiamide (10%) and was placed into small Perspex columns (25 mm in diameter). Nitrogen or air was either flushed through the column at a flow rate of 60 ml min^{-1}, or across the headspace of the column at a flow rate of 40 ml min^{-1}. The NO flux was calculated from the difference in NO concentration at the column inlet and outlet, measured by chemiluminescence. Vertical flow should ensure that the bulk of the NO produced is carried away with the air stream, before undergoing further reduction, whereas horizontal flow over the surface of the column will mostly transport the NO emitted from the soil surface. Under aerobic and anaerobic conditions, only 13% of the NO produced was emitted from the soil surface (Figure 7). In spite of adding nitrification inhibitor and using NO_3^- as the N source, NO was produced under fully aerated conditions; however, this was only at a fifth of the rate under anaerobic conditions. The most likely explanation for this is the presence of anaerobic microsites, created by soil clumping in this experiment.

Soil Emissions of NO and N_2O

The magnitude of NO and N_2O emissions from soils depends primarily on the

72

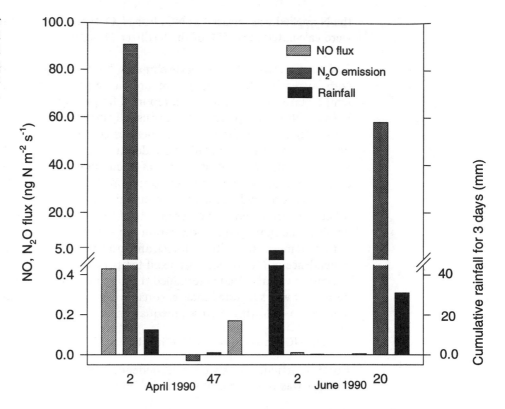

Figure 8 The effect of fertilizer application and rainfall and the emissions of NO and N_2O from clay loam soils cropped with ryegrass cut for silage in South Scotland.

Days after application of NH_4NO_3 (100 kg N ha^{-1})

concentration of N in the soil, the soil temperature, bulk density, and water content. The characteristics of the water content primarily regulate the aeration status of the soil, thus determining the transformation pathway (nitrification and denitrification) and its rate.

Soil Nitrogen Status. Of all the parameters listed above, the soil N status is the most important variable determining the magnitude of the NO and N_2O emission rates. Agricultural soils, immediately after the application of N fertilizer, are therefore found to emit NO and N_2O at the highest rates. The fertilizer effect usually lasts for 2–3 weeks, and relative rates of NO and N_2O emissions within this time period depend on the rainfall and wetness of the soil. For example, high N_2O and low NO emissions were measured immediately after the spring application of N fertilizer to ryegrass (cut for silage), when rainfall was frequent and the soil was wet. In June, however, when N fertilizer was applied during a very dry period, no immediate fertilizer response in N_2O emission was observed, but the NO emission was an order of magnitude greater than in April. Frequent rain events 2 weeks after the summer fertilizer application triggered the emission of N_2O, but stopped the emission of NO (Figure 8). From a wide range of NO emission studies in temperate climates it was estimated that, on average, 0.26% of the N applied was released as NO. For N_2O, typical fertilizer-induced emissions were calculated at 0.95% of the fertilizer N applied.[51]

Soil Temperature. In temperate climates, NO and N_2O emission rates increase with increasing soil temperature and a response to diurnal and seasonal temperature variations has been reported frequently.[4,52] Activation energies for both soil NO and N_2O emissions are usually in the range of $30–150\,kJ\,mol^{-1}$.[53,54] This range matches the activation energies quoted for many other soil microbial processes, for example nitrification, denitrification rates, and methane emission and oxidation rates. The usefulness of calculating activation energies as a tool in determining processes is therefore questionable. However, it has been suggested that unreasonably large activation energies indicate the greater importance of other soil parameters that determine the flux and can also be used to deduce the depth of the main production within the soil profile.[55]

For a range of agricultural, forest, and moorland soils in southern Scotland, the dependence of NO emissions on soil temperature changes was the second most important variable that determined the magnitude of the emission.[56] Using this data set, it was suggested that reasonable estimates of NO emission rates can be made from following regression equation:

$$\log (\text{NO emission}) = -3.23 + 1.01 \log (\text{soil NO}_3) + 0.165 (\text{soil temperature}).$$

For this particular data set, the dependence of N_2O emissions on soil temperature changes was not as conclusive as for NO.

Soil Water Content. The aeration status of the soil is closely regulated by the proportion of the pore volume occupied by water, thus determining the rates of NO and N_2O emissions. In particular for N_2O, changes in soil water content are very important in determining the emission rate. A model of the relationship between water-filled pore space (WFPS) and gaseous N emissions from soil suggests that no N_2O is emitted when the WFPS is less than 30%, with maximum emissions occurring between 50% and 70%. Nitric oxide emissions are much less responsive to changes in WFPS, with near maximum to maximum emissions occurring over a much wider range (approx. 25–60%).[57] Generally, field measurements have shown an increase in N_2O emissions with increasing soil water content, rainfall, *etc.*, whereas NO emissions decrease under the same conditions (Figure 8).[4]

[51] I. P. McTaggart, H. Clayton and K. A. Smith, in *Non-CO₂ Greenhouse Gases*, ed. J. van Ham, L. J. H. M. Jannsen and R. J. Swart, Kluwer, Dordrecht, 1994, p. 421.

[52] K. A. Smith, H. Clayton, I. P. McTaggart, P. E. Thomson, J. R. M. Arah and A. Scott, *Philos. Trans. R. Soc. London, Ser. A.*, 1995, **351**, 327.

[53] F. Slemr and W. Seiler, *J. Atmos. Chem.*, 1984, **2**, 1.

[54] R. Conrad, W. Seiler and G. Bunse, *J. Geophys. Res.*, 1983, **88**, 6709.

[55] R. Conrad and W. Seiler, *Soil Biol. Biochem.*, 1985, **17**, 893.

[56] U. Skiba, D. Fowler and K. A. Smith, in *Non-CO₂ Greenhouse Gases*, ed. J. van Ham, L. J. H. M. Jannsen and R. J. Swart, Kluwer, Dordrecht, 1994, p. 153.

[57] E. A. Davidson, in *Production and Consumption of Greenhouse Gases: Methane, Nitrogen Oxides, and Halomethanes*, ed. J. E. Rogers and W. B. Whitman, American Society for Microbiology, Washington, 1991, p. 219.

Field Measurements of NO and N_2O Exchange

The large number of different soil properties which control the emission rates of NO and N_2O lead to great spatial variability in emissions, even at scales as small as 0.1 m. Prediction of landscape scale fluxes from measurements made using soil cores or chambers is, therefore, a very speculative exercise and extremely prone to error. Use of micrometeorological methods enables fluxes to be averaged over areas of 10^2 to $10^5 \, m^2$, which integrates the small-scale (within field) spatial variability in emissions and reduces these scaling problems. The methods allow both emission and deposition fluxes to be measured and provide methods of estimating the net exchange of a range of gases to be studied.

The Halvergate Experiment. The trace nitrogen gases exchanged between agricultural land and the atmosphere are generally studied in individual compound-specific experiments. In an attempt to quantify the relative magnitudes of a range of trace gas fluxes [NO, NO_2, HNO_3, N_2O, NH_3, PAN (peroxyacetyl nitrate) and HONO], a large field experiment over grassland at the Halvergate marshes was conducted during September 1989.[58] In addition to the range of gases, a variety of methods including enclosures of three different kinds, micrometeorological methods using both eddy covariance and flux-gradient and energy balance methods were used.

The site was a drained marsh which received no artificial N inputs, although cattle were present on the site until a couple of weeks before the experiment. N_2O emissions were measured by chamber techniques as no instrumental techniques were sensitive enough at that stage to permit micrometeorological measurements. Although spatially very variable, the mean emission rate from the site was 4 ng N_2O-N$m^{-2}s^{-1}$. The sporadic measurements made impossible the determination of any response to temperature or water status.

Emissions of NO varied between 5 and 20 ng $N m^{-2}s^{-1}$ (Figure 9), but again no effect of temperature was observed. NO_2 was deposited to the vegetation by dry deposition (Figure 9) at rates up to 30 ng $N m^{-2}s^{-1}$, and this process was shown to be controlled by stomatal resistance. Figure 10 shows canopy resistances calculated both for NO_2 and O_3, illustrating that the same physical process controls deposition rates for both gases. The effect of early stomatal closure due to water stress is also apparent after 12:00 GMT.

Although the peak deposition rate of NO_2 is greater than the maximum emission rate of NO, estimates of the long-term mean exchange rates must take into account the diurnal cycle of NO_2 uptake and the dependence of uptake rate upon atmospheric NO_2 concentration. The estimated net summer fluxes of these species at Halvergate, taking the diurnal changes in canopy resistance to NO_2 deposition into account, are 10 ng NO-N$m^{-2}s^{-1}$ emission and 5 ng NO_2-N$m^{-2}s^{-1}$ deposition.

Fluxes of HNO_3 and HONO were measured by the flux-gradient technique,[59] yielding 19 separate estimates of the fluxes. The mean deposition rate for HNO_3 was 5.5 ng $N m^{-2}s^{-1}$ and all the observed fluxes were towards the ground. For

[58] K. J. Hargreaves, D. Fowler, R. L. Storeton-West, and J. H. Duyzer, *Environ. Pollut.*, 1992, **75**, 53.
[59] R. M. Harrison and A.-M. N. Kitto, *Atmos. Environ.*, 1994, **28**, 1089.

Figure 9 The emission and deposition of NO and NO$_2$ at the Halvergate site on 15 September 1989. By convention, a negative flux indicates deposition.

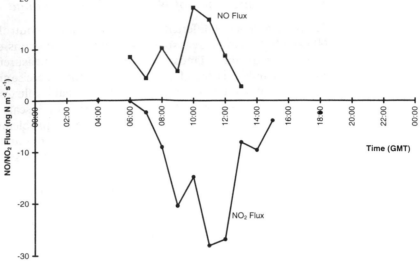

Figure 10 Canopy resistances for NO$_2$ and O$_3$ deposition at the Halvergate site on 15 September 1989. The resistances show that deposition is controlled by stomatal conductance, and the effect of early afternoon stomatal closure is illustrated.

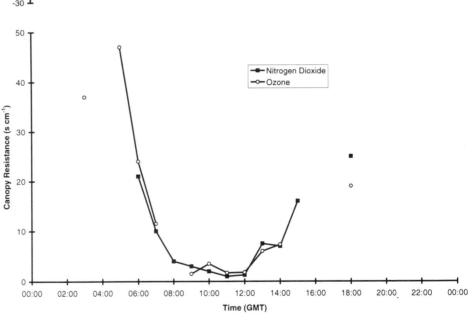

HONO the mean flux was an emission of 1 ng N m^{-2} s^{-1}, but this includes periods both of emission and deposition. On several occasions, no concentration gradients were detected. The direction of the flux was dependent on NO$_2$ concentration, with emission observed only when NO$_2$ concentration was less than 10 ppb. The process of HONO exchange appears to be regulated by the net result of small deposition flux to the surface and a surface chemistry production of HONO from NO$_2$. Fluxes of PAN deposition were measured using a chamber technique[60] and were small (less than 0.5 ng N m^{-2} s^{-1}).

[60] G. J. Dollard, B. M. R. Jones and T. J. Davies, *Dry Deposition of HNO$_3$ and PAN*, AERE Harwell Report R13870, Harwell Laboratory, Didcot, 1990.

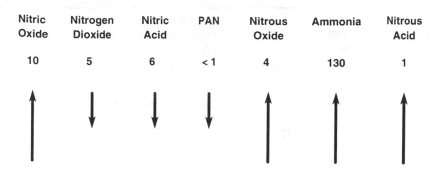

Figure 11 A budget for gas-phase trace nitrogen species at Halvergate, September 1989.

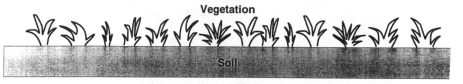

The land–atmosphere fluxes of individual trace gases over the Halvergate grassland are summarized in Figure 11. The data reveal the range of individual gases contributing to the net exchange and the dominance of NH_3 fluxes at this site. It is unusual to have such a large range of measurements at a single site, but fluxes of these gases are present over all agricultural land, and it is the magnitude and direction of the individual terms which present a challenge. So far, the primary controls regulating exchange of these gases have been identified for most components (NO, NO_2, HNO_3, N_2O, and NH_3) and are summarized in Table 4, but the soil, plant, atmospheric and air chemistry data to enable long-term regional fluxes to be determined are lacking for most components. Expressed as a daily flux, it was estimated[61] that the site was losing approximately 115 g $N\,ha^{-1}\,d^{-1}$ (Sutton, unpublished data).

Up-scaling N_2O and NO Fluxes to the Field Scale

The field experiments at Halvergate illustrate the range of gases contributing to the net exchange of nitrogen over agricultural land. Another problem in up-scaling fluxes is the fine-scale spatial variability in emissions of gases which are regulated by soil physical properties, as well as soil NO_3^- and temperature. These are, of course, the NO and N_2O fluxes under microbiological control.

[61] D. Fowler, J. N. Cape, M. A. Sutton, R. Mourne, K. J. Hargreaves, J. H. Duyzer and M. W. Gallagher, in *Acidification Research: Evaluation and Policy Applications*, ed. T. Schneider, Elsevier, Amsterdam, 1992.

Table 4 Primary processes controlling land–atmosphere exchange of gases

Trace gas	Primary processes regulating land–atmosphere exchange (secondary)	Availability of data and including models to calculate annual fluxes	
		Data	Models
NO_2	Stomatal resistance (turbulent transfer processes)	√	√
HNO_3	Turbulent transfer processes	HNO_3 data lacking	√
NO	Soil nitrogen, soil temperature, soil water (soil physical properties)	√ approximately	√ empirical
NH_3	Canopy compensation point, ambient NH_3 concentration, surface water (SO_2 concentration)	For limited areas	√
N_2O	Soil nitrogen, soil temperature, soil water (soil physical properties)	√ approx. only	√ empirical

In the case of N_2O fluxes, the techniques which have provided most measurements to date have relied on enclosures within which the accumulation of N_2O with time provides a measure of emission. In most cases, the area of measurement is between 0.1 and $1.0\,m^2$, yet spatial variability in emission over a field may be of several orders of magnitude over distances of a few metres. A field experiment was carried out at Stirling, central Scotland, in April 1992 to measure N_2O emission rates from a grazed, fertilized pasture and to investigate the processes of scaling up fluxes measured by chambers to those determined by micrometeorology.[62,63] Immediately before the experiment began, an application of $185\,kg\ NH_4NO_3\,ha^{-1}$ was made. Fluxes were measured over a two-week period using a wide variety of chamber and micrometeorological techniques, and the results illustrate the problems to be overcome in scaling up from measurements over small areas of less than $1\,m^2$ to the field scale.

The emissions of NO from soil may be oxidized readily within plant canopies to NO_2, which may then be absorbed by stomata within the canopy or emitted from the canopy to the atmosphere. These processes, described by Pilegaard *et*

[62] K. J. Hargreaves, U. Skiba, D. Fowler, J. R. M. Arah, F. G. Wienhold, L. Klemedtsson and B. Galle, *J. Geophys. Res.,* 1994, **99**, 16 569.
[63] K. A. Smith, H. Clayton, J. R. M. Arah, S. Christensen, P. Ambus, D. Fowler, K. J. Hargreaves, U. Skiba, G. W. Harris, F. G. Wienhold, L. Klemedtsson and B. Galle, *J. Geophys. Res.,* 1994, **99**, 16 541.

al.,[64] may represent an important nitrogen cycling mechanism within the soil–plant canopy, and are illustrated diagrammatically in Figure 12.

The three micrometeorological methods [Fourier Transform Infrared Spectroscopy (FTIR), Tunable diode laser spectroscopy (TDL), and gas chromatography (GC)] all gave measurements over essentially the same area of the field using the flux-gradient technique, and Figure 13 shows the good agreement obtained between the methods, with emission fluxes in the range 0–140 ng $N m^{-2} s^{-1}$. However, the chamber results also shown on Figure 13 gave much larger fluxes of 140–350 ng $N m^{-2} s^{-1}$, a difference which is hard to explain in terms of systematic differences between methods. There was strong evidence that the difference results from the large spatial variability in N_2O emission and the associated 'hot-spots' in the field which lay outside the 'footprint' of the micrometeorological equipment.[63] The experiment demonstrates that a variety of micrometeorological methods may all yield broadly similar fluxes, and that these average fluxes are at the field scale. However, the comparison with chamber methods illustrates some of the difficulties in comparisons at different scales. In the case of chamber measurements, it is necessary to use large numbers of chambers to overcome spatial heterogeneity in emissions, yet to compare with the micrometeorological methods correctly the footprint of the field in which the chambers are placed (which varies with wind direction, speed and atmospheric stability) must be integrated in the same way for both techniques. In practice, this is an extremely demanding task which, for emissions of N_2O and NO, may only be achieved within a substantial range of uncertainty (± 20–25%). As such, the data in Figure 13 represent progress but with more chambers (or a more fortunate wind direction) could have been better by a factor of 2.

NO emissions did not exceed 2 ng $N m^{-2} s^{-1}$ and their measurement was only possible by chamber methods. The low NO emissions but high N_2O emissions show that denitrification was the main source of N_2O at this site. The discrepancies between the chamber and micrometeorological methods illustrated the need to define the flux-footprint of a micrometeorological measurement very carefully, and to use this information in the field to choose the locations in which chambers are placed. Without such an approach, the integration of results from chambers into estimates of field-scale emission remains an uncertain method.

The contrast between Stirling and Halvergate is apparent in the large difference in $NO:N_2O$ emission ratio, and reflects the effects of applying large N fertilizer inputs to a heavy clay soil (Stirling), thus (in spite of the low soil temperatures) providing better conditions for denitrification to occur than in the drained, aerated soil (Halvergate).

Upscaling Annual NO and N_2O Emissions to Regions ($\geqslant 100 km \times 100 km$)

Williams *et al.*[47,48] produced an inventory of soil NO emissions for the United States based on land use category and soil temperature variations, and estimated that on an annual basis the soil source was only 6% of the man-made NO_x source.

[64] K. N. Pilegaard, N. O. Jensen and P. Hummelshoj, in *Acid Rain Research: Do We Have Enough Answers?*, ed. G. J. Heij and J. W. Erisman, Elsevier, Amsterdam, 1995, p. 502.

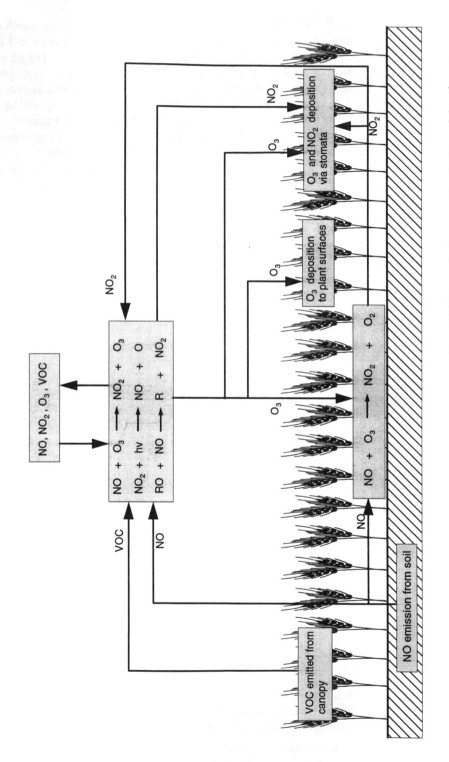

Figure 12 Interactions of soil emissions of NO with O_3 in plant canopies and NO_2 uptake by vegetation in determining the net exchange of NO_x between soil–plant and the atmosphere.

Figure 13 N$_2$O emission from the Stirling site in April 1992. FTIR, GC, and TDL plots are micrometeorological methods. The chambers averaged over 0.126 m^2 and 0.49 m^2.

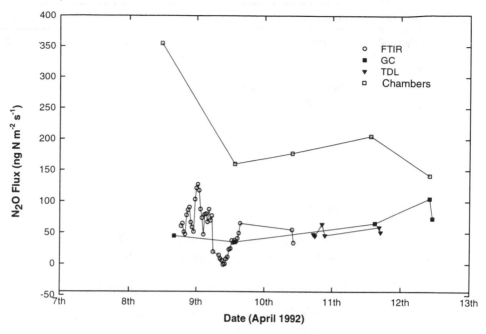

By adopting a similar very simplistic approach for Europe, the calculated NO emission and yields are similar. Estimates of NO emission in Europe by Simpson[65] suggest that, for most countries, soil emissions make up 5–10% of the total NO$_x$ emissions, but this can be much larger where man-made emissions are small, *e.g.* Turkey where 77% of the total NO$_x$ emissions are soil derived. For the UK, annual soil NO emissions were estimated at 23 kt N, which is 2.8% of the annual man-made emissions. A slightly different approach is taken here, where soil NO emission rates published for temperate climates were categorized into 'recently fertilized tilled land', 'not recently fertilized tilled land', 'grassland' or 'forests' (48 observations) and used to estimate emissions for the land classes in the UK listed in Table 5. It is encouraging to see that the estimate of annual soil NO emissions (20 kt N) is very similar to Simpson's estimate. Agricultural land is undoubtedly the main source of soil NO, being responsible for over 80% of the total soil NO emissions.

For N$_2$O, estimates of soil emissions are perhaps slightly more straightforward, simply because so much more information on soil N$_2$O emissions is available. For agricultural soils, estimates as a percentage of fertilizer input have been published[65,51] and, for the remaining land use classes listed in Table 5, mean annual emissions from a series of long-term flux measurements are available.[66] The total annual soil N$_2$O emissions for the UK (27 kt N) are very similar to the annual soil NO emissions. Soils, however, have by far a greater impact on the

[65] D. Simpson, *Biogenic VOC Emissions in Europe: Modelling the Implications for Ozone Control Strategies*, AWMA Conference, San Diego, California, 1993.

[66] U. Skiba, I. P. McTaggart, K. A. Smith, K. J. Hargreaves and D. Fowler, *Energy Convers. Manage.*, 1996, **37**, 1303.

Table 5 Estimates of soil N_2O and NO emissions in the UK

Land cover class	km^2	kt N_2O-N yr^{-1}	kt NO-N yr^{-1}
Managed grasslands	67 690	19.0	8.6
Tilled land	41 760	4.7	8.4
Deciduous woodland	12 300	1.2	0.3
Coniferous woodland	13 700	0.5	0.4
Rough grass, moorland grass	32 700	0.8	0.9
Heath/bog/bracken	39 300	1.3	1.0
Others	26 600	Not determined	Not determined
Total		27	20

total UK N_2O (30%) than NO_x emissions (3%) and will become the main source of atmospheric N_2O when adipic production in the UK ceases. Agricultural grasslands are the main source of soil N_2O emissions, being responsible for around 60% of the total annual soil emissions.

Global N_2O Emission and Agriculture

The agricultural emissions of N_2O have been recognized as contributors to a loss of fertilizer for many years, but only recently has the global significance of the agricultural N_2O emission been treated as an important global issue.[67]

The presence of N_2O in the atmosphere contributes to the radiative forcing of global climate and to stratospheric ozone chemistry.[68] The destruction of N_2O in the stratosphere is believed to be quite well defined at 8.7 Tg N yr^{-1}.[69] The rate of accumulation of N_2O in the atmosphere (0.25%) is equivalent to about 4 Tg N yr^{-1} and thus the atmospheric lifetime is about 170 years. Estimates of the pre-industrial atmospheric N_2O concentration from ice core analysis allow the natural emission to be estimated at about 8 Tg N y^{-1}; thus the anthropogenic source is around 5 Tg N yr^{-1} (summarized in Table 6). The primary sources are soils for both natural and anthropogenic emission. The Intergovernmental Panel on Climate Change (IPCC) assessment of N_2O emissions has recently recognized the importance of cultivated soils in the global N_2O budget, increasing the global contribution of N_2O from a range of 0.01–2.2 Tg N yr^{-1} in the 1990 report to the current, much longer and apparently more certain range of 1.8–5.3 Tg N yr^{-1} (Table 7). The likely N_2O flux from agriculture of about 4 Tg N yr^{-1} represents the majority of the anthropogenic flux and most of the annual accumulation in the atmosphere. It follows that if measures are ever developed to regulate these emissions the agricultural industry will be a primary target.

4   Conclusions

The gaseous losses of nitrogen from agriculture include a range of oxidized (NO and N_2O) as well as reduced (NH_3) compounds. These gases form a part of the

[67] IPCC, *Climate Change 1994*, Intergovernmental Panel on Climate Change, Cambridge University Press, 1994, p. 339.

[68] R. P. Wayne, *Chemistry of Atmospheres*, Oxford Scientific, Oxford, 1991, p. 447.

[69] T. E. Graedel and P. J. Crutzen, *Atmospheric Change*, Freeman, New York, 1993, p. 446.

ble 6 Estimated sources of nitrous oxide ($TG\,N\,yr^{-1}$) (IPCC)[67]

Source	IPCC, 1994*	IPCC, 1990
Oceans	1–5	1.4–2.6
Tropical soils		
wet forests	2.2–3.7	2.2–3.7
dry savannas	0.5–2.0	
Temperate soils		
forests	0.1–2.0	0.7–1.5
grasslands	0.5–2.0	
Cultivated soils	1.8–5.3	0.01–2.2
Biomass burning	0.2–1.0	0.02–0.2
Industrial sources	0.7–1.8	
Cattle and feed lots	0.2–0.5	
Total identified sources	7–23	4.4–10.2
Total emissions from soils	5–15 = 65–70% of total emissions	

*IPCC = Intergovernmental Panel on Climate Change.

Table 7 Global sources of nitrous oxide*

Global N_2O sources	$Tg\,N\,yr^{-1}$	
Stratospheric distribution of N_2O	8.7	sinks
Atmospheric accumulation of N_2O (0.25% yr^{-1})	4.0	
Natural emissions (largely from tropical soils)	8.0	sources
Anthropogenic emissions cultivated soils (+industry and combustion)	4.7	

*Adapted from Graedel and Crutzen, 1993,[69] and IPCC, 1994[67].

natural biogeochemical cycling of nitrogen, but in the case of heavily fertilized land the fluxes are larger, in some cases by orders of magnitude, than those in natural ecosystems. There is also a tendency for some cycling of gaseous nitrogen to occur within canopies of semi-natural vegetation, while in the case of agricultural crops the fluxes generally represent emission.

The net losses of fixed nitrogen from UK agriculture are dominated by the emission of NH_3 from cattle, poultry and pig production. The loss of nitrogen by volatilization of NH_3 in the UK amounts to approximately 350 kt N annually and is equivalent to about 25% of the inorganic fertilizer applied annually. The reactive nature of atmospheric NH_3, and its scavenging by precipitation and rapid deposition onto vegetation, leads to a substantial deposition of NH_3 within the country. The total (wet plus dry) deposition of NH_x-N in the UK is 230 kt N annually, and represents about 69% of emissions. The environmental problems caused by NH_3 deposition within the UK are, therefore, largely a consequence of UK emissions rather than emission elsewhere in Europe. However, a significant quantity (approx. 34%) of reduced nitrogen from the UK is 'exported' by the wind to other parts of Europe, largely as particulate $(NH_4)_2SO_4$ and NH_4NO_3.

Emissions of oxidized nitrogen from UK soils, in contrast with those of agricultural NH_x emission, are much smaller and amount to 20 kt NO-N and

27 kt N_2O-N annually. These emissions, while of little practical consequence locally, represent one of the major causes of increased N_2O emissions globally. The N_2O makes an important contribution to radiative forcing, contributes to stratospheric ozone chemistry and has a very long atmospheric lifetime (150 y); accordingly the relatively small annual emissions of N_2O from UK soils should not be overlooked. Emissions of NO from soil represent only a few percent of UK emissions from combustion processes and vehicles. While the total NO emissions are a major national and international issue, especially for the production of photochemical oxidants, the major interest in soil emissions of NO is in the global NO emission.

The brief consideration of deposition as well as emission processes was included to highlight the interaction between emissions and deposition of reactive nitrogen, and the influence land use has on the net exchange. In considering the influence of agriculture on emissions of trace nitrogen gases, it is also important to consider the effect of land use on the sink strength for both oxidized and reduced nitrogen.

Drugs and Dietary Additives, Their Use in Animal Production and Potential Environmental Consequences

THOMAS ACAMOVIC AND COLIN S. STEWART

1 Introduction

The title of this article is extremely wide-reaching in the subject matter that is embraced. The complexity and number of relevant compounds is extremely large and these can be man-made or may be produced by natural organisms, including plants and microbes. The various aspects that can be considered are the chemistry, biochemistry and microbiology of the compounds, their effects in animals, plants, and microbes, as well as their effects on the environment. The usage and regulations that govern the use of drugs vary widely throughout the world, as do the methods by which such compounds are administered. There is a large variety of animals which are reared for a variety of purposes, including food production; these include fish, reptiles, small rodents, chickens, turkeys, sheep, goats, cattle, deer, horses, ostriches, kangaroos and other less well appreciated species.[1]

Within the constraints of this article it is impossible to be comprehensive in the coverage of the subject matter, in terms of the chemicals involved and in the widely varying practices and areas of the world in which the title compounds are ingested by farmed animals. This account is, however, intended to give an overview, citing some relevant examples, of the beneficial and adverse effects, in animals and on the environment, of man-made compounds and 'naturally' produced compounds in extensive and commercial production systems.

The definition of a drug differs between dictionaries and among the various professional specialisms. A search of the internet elicited various definitions and a paraphrase of the most memorable is 'a compound can be defined as a drug if, when injected into a rodent, it yields a scientific publication'. Although this is a memorable definition, for the purposes of this review, however, a drug is defined broadly as a compound that has properties that influence the health of an animal when ingested or administered to that animal. A brief look at current literature will quickly convince the reader that this is a definition which covers man-made and natural compounds that can be extracted from plant material and microbes and used.[2]

---

[1] W.C. Campbell, *Ivermectin and Abermectin*, Springer, New York, USA, 1989.

It should be noted that such a definition does not necessarily mean that drugs are always used as medicines to promote good health, not does it imply that drugs are man-made compounds that are administered to an animal by a veterinary or other practitioner. Drugs, even when these are administered for medicinal purposes to animals, can have adverse effects on animal and human health, especially if excessive amounts are given.[2-4] The effect can vary in severity, depending on the animal species, quantity consumed/administered, route of administration, health and nutritional status of the animal. Drugs and other dietary additives can also have beneficial and adverse environmental effects. The use of some of the waste products from animals, which can be of high protein content with high digestibility, *e.g.* excreta and offal, as feed supplements for other animals[5-7] is considerably influenced by the presence of absence of drugs, since their presence in such material may allow deposition in tissue and other products destined for the human food market.[8,39,44]

We shall cover compounds with which animals come into contact because of their inclusion in diets and their administration by man, as well as those that the animals encounter as a consequence of their natural feeding regimes.

2 Commercial Implications of the Use of Drugs and Dietary Additives

At the time of preparation of this article the consumption of meat in the UK for 1995 and 1996 was estimated to be about $3.6\,Mt\,yr^{-1}$;[9] thus there is a considerable demand to produce it economically with minimal adverse impact on the environment by maintaining good animal health and welfare.

Thus the requirement for the use of man-made drugs and dietary additives as veterinary medicines for the treatment of farmed animals is considerable and worth about 100 million pounds sterling annually in the UK (£104 million in 1994).[10] The investment in dietary additives such as vitamins, trace minerals, coccidiostats, pigmenters, enzymes and other probiotics to feed compounders in the UK is worth about £110 million, assuming an addition rate of 2.5 kg per tonne and a cost of approximately 3% of the total concentrate dietary cost (calculated from MAFF data, 1995).[10] These data can be increased by a factor of about 10 when the compound feed produced within Europe is considered.[10,11]

2   Y. Debuf, *The Veterinary Formulary Handbook of Medicines Used in Veterinary Practice*, The Pharmaceutical Press, London, 2nd edn., 1994.

3   J. N. Hathcock, D. G. Hattan, M. Y. Jenkins and J. T. McDonald, *Am. J. Clin. Nutr.*, 1990, **52**, 183.

4   H. Nau, in *Developments and Ethical Considerations in Toxicology*, ed. M. I. Weitzner, The Royal Society of Chemistry, Cambridge, 1993, p. 35.

5   A. Donkoh, P. J. Moughan and W. C. Smith, *Anim. Feed Sci. Technol.*, 1994, **49**, 57.

6   S. P. Rose, D. M. Anderson and M. B. White, *Anim. Feed Sci. Technol.*, 1994, **49**, 263.

7   A. R. Patil, A. L. Goetsh, B. Kouakou, D. L. Galloway Sr, L. A. Forster Jr and K. K. Park, *Anim. Feed Sci. Technol.*, 1995, **55**, 87.

8   R. J. Heitzman, *Veterinary Drug Residues. Residues in Food Producing Animals and their Products: Reference Materials and Methods*, Final Report EUR 14126EN, Office for Official Publications of the European Communities; Luxembourg, 1992.

9   MLC, *Meat Demand Trends*, Meat and Livestock Commission, 95/3, Milton Keynes, August 1995.

10 MAFF, *Agriculture in The United Kingdom*, HMSO, London, 1995.

11 *Proceedings of the 2nd European Symposium on Feed Enzymes*, Noordwijkerhout, Netherlands, ed. W. van Hartingsveldt, M. Hessing, J. P. van der Lugt and W. A. C. Somers, 1995.

The ingestion of natural compounds, some of which can be identical to those administered for medicinal purposes,[12,13] which have adverse effects especially in extensively reared animals, has been estimated to be about the equivalent of a 2% tax on livestock production.[14] For example, in the UK this would amount to about £20 million (calculated from MAFF data, 1995).[10]

3 Compounds Encountered by Animals

An investigation of the literature confirms that drugs can range widely in their chemical nature, from relatively innocuous naturally produced vegetable oils and tannins to synthetic surfactants such as Poloxalene, Dimethicone, and the inorganic salt Na_2CO_3, which are all used to treat bloat (gases trapped in the gastrointestinal tract) in ruminants.[2,15–17] Other examples of compounds which can be classified as drugs include vitamin E and other vitamin and trace element supplements that can be administered in the diet and by injection.[2,4,16] Examples of other compounds encountered by animals are given in Figure 1.

Some of these compounds could be considered as dietary additives, but various other terms, including pesticides,[1] can also be used. They can have beneficial effects on the environment and this aspect will be discussed later. The ionophore monensin, which is an alicyclic polyether (Figure 1), is a secondary metabolite of *Streptomyces* and aids the prevention of coccidiosis in poultry. Monensin is used as a growth promoter in cattle and also to decrease methane production, but it is toxic to equine animals.[2,16,18,19] Its ability to act as an ionophore is dependent on its cyclic chelating effect on metal ions.[20] The hormones bovine somatotropin (BST) and porcine somatotropin (PST), both of which are polypeptides, occur naturally in lactating cattle and pigs, respectively, but can also be produced synthetically using recombinant DNA methods and administered to such animals in order to increase milk yields and lean meat production.[21–26]

[12] A. A. Seawright, M. P. Hegarty, L. F. James and R. F. Keeler, *Plant Toxicology*, Queensland Dept of Primary Industries, Yeerongpilly, Australia, 1985.

[13] P. R. Cheeke and L. R. Shull, *Natural Toxicants in Feeds and Poisonous Plants*, AVI, Westport, CT, 1985.

[14] L. F. James, in *Plant Associated Toxins*, ed. S. M. Colegate and P. R. Dorling, CAB International, Wallingford, 1994, p. 1.

[15] G. C. Waghorn and W. T. Jones, *N. Z. J. Agric. Res.*, 1989, **32**, 227.

[16] G. Walker, *Compendium of Data Sheets for Veterinary Products*, Datapharm, Enfield, 1994.

[17] V. Parbhoo, H. R. Grimmer, A. Cameron-Clarke and R. M. McGrath, *J. Sci. Food Agric.*, 1995, **69**, 247.

[18] R. J. Wallace, in *Proceedings of 8th Annual Symposium of Biotechnology in the Feed Industry*, ed. T. P. Lyons, Alltech Technical Publications, Nicholasville, 1992, p. 193.

[19] K. A. Johnson and D. E. Johnson, *J. Anim. Sci.*, 1995, **8**, 2483.

[20] B. C. Pressman, *Annu. Rev. Pharmacol.*, 1976, p. 501.

[21] D. E. Bauman, S. N. McCutcheon, W. D. Steinhour, P. J. Eppard and S. J. Sechen, *J. Anim. Sci.*, 1985, **60**, 583.

[22] M. Vanbelle, in *Proceedings of 5th Annual Symposium of Biotechnology in the Feed Industry*, ed. T. P. Lyons, Alltech Technical Publications, Nicholasville, 1989, p. 191.

[23] A. N. Pell, D. S. Tsang, B. A. Howlett, M. T. Huyler, V. K. Meserole, W. A. Samuels, G. F. Hartnell and R. L. Hintz, *J. Dairy Sci.*, 1992, **75**, 3416.

[24] D. E. Bauman and R. G. Vernon, *Annu. Rev. Nutr.*, 1993, **13**, 437.

[25] D. Schamms, in *Ruminant Physiology: Digestion, Metabolism, Growth and Reproduction*, ed. W. v. Engelhardt, S. Leonhard-Marek, G. Breves and G. Giesecke, Ferdinand Enke, Stuttgart, 1995, p. 429.

[26] F. R. Dunshea, Y. R. Boisclair, D. E. Bauman and A. W. Bell, *J. Anim. Sci.*, 1995, **73**, 2263.

Figure 1 Compounds encountered by animals

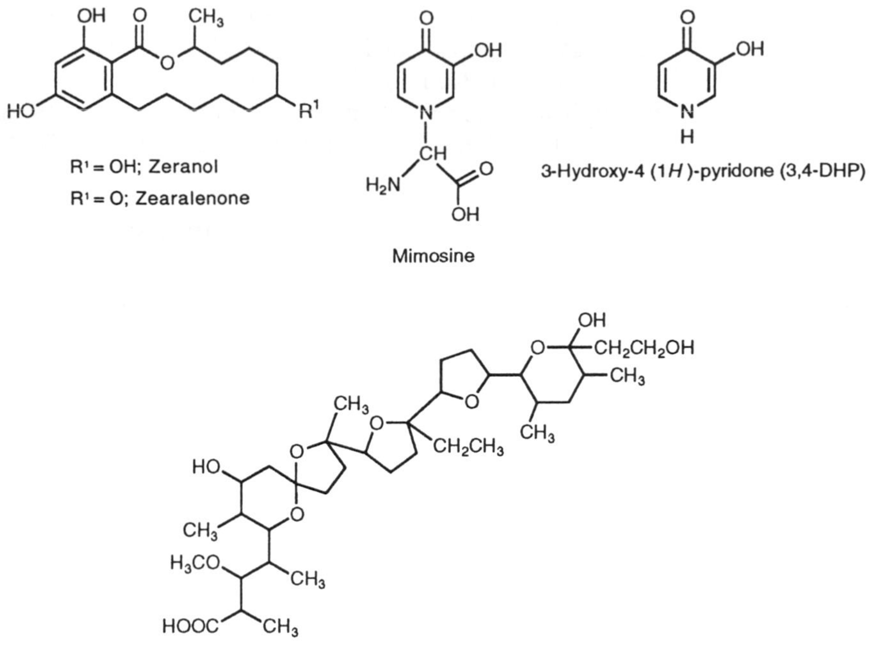

R¹ = OH; Zeranol

R¹ = O; Zearalenone

Mimosine

3-Hydroxy-4 (1*H*)-pyridone (3,4-DHP)

Monensin

Considerable controversy exists regarding the use of BST and PST in animal production systems, in spite of increased efficiencies of production, because of the perceived ethical and human health problems.[22,27] In some cases, drugs which are licenced for use in some animal species are used illegally for other animals, with possible adverse effects on human health. Spectinomycin, which is an antibiotic approved in the USA for use in poultry, is an example of a drug which has been used without approval in lactating cattle, with the consequence that residues are deposited in the milk and meat.[2,28] Gentamicin, which also is an antibiotic, is not approved in the USA by the Food and Drug Administration but can be administered when prescribed by a licensed vet. In the UK it is a prescription-only medicine (POM), which can be administered in the diet or by injection, but the biological half-life varies depending on the method of administration and the species of animal to which the material has been administered.[2,16,29-31] When such drugs are administered, it is obligatory for a withdrawal period to be exercised prior to utilizing the products for human consumption.

Animals which are fed extensively can encounter a vast array of compounds of varying medicinal and toxic potency and susceptibilities to degradation or detoxification; a number of texts can be consulted.[2,13,32-38]

27 D. E. Bauman, *J. Dairy Sci.*, 1992, **75**, 3432.
28 P. G. Schermerhorn, P. S. Chu and P. J. Kijak, *J. Agric. Food Chem.*, 1995, **43**, 2122.
29 S. A. Brown and N. Baird, *Am. J. Vet. Res.*, 1988, **49**, 2056.
30 M. A. Fennel, C. E. Uboh, R. W. Sweeney and L. R. Soma, *J. Agric. Food Chem.*, 1995, **43**, 1849.
31 S. K. Garg, S. K. Verma and R. P. Uppal, *Br. Vet. J.*, 1995, **151**, 453.

The public perception of the use of drugs and other dietary additives in animal production, and the presence of these in the various products, varies throughout the world.[8,39,44] There is a demand for drug- and pesticide-free produce. This was supported recently by a sign in a supermarket in Zimbabwe where meat for sale was defined as 'containing no added poisons'.[40] There is also a demand in different areas of the world for meat and other products to be a suitable creamy or yellow colour which, frequently in intensively reared animals, requires the addition of man-made (sometimes defined as naturally identical) carotenoid pigments to animal diets.[41–43] The yellow pigmentation due to carotenoids may also have beneficial effects on the animals and the consumers of meat, milk, and eggs because of their antioxidant and anticarcinogenic properties.[43]

Thus a drug can be man-made or can arise as a secondary metabolite in plants or microbes and drugs are often classified and registered accordingly. Some man-made and natural drugs are presented in Figure 1 and Table 2. It should be noted that there is often a close similarity between the structures of man-made and naturally produced compounds. This should not be surprising since some of the compounds of interest, such as BST, monensin and spectinomycin, are produced by both natural and genetically engineered organisms.[16,18,22,26,27] The content of such compounds, including those that have originated naturally, in various animal feeds is often required to be specified by law, but the requirement varies between communities and governments around the world.[8,39,44] Growth hormones have been shown to improve growth and efficiency in animals but their use, either as dietary supplements or as implants, has been discouraged in some areas of the world and meat products obtained in this manner may not be sold in such restricted markets. Similarly, the use of synthetically produced hormones such as PST and BST causes restriction in the use of milk in various markets.[8,22,39,44–48]

32 J. P. F. D'Mello, *Anim. Feed Sci. Technol.*, 1992, **38**, 237.

33 J. P. F. D'Mello, C. M. Duffus and J. H. Duffus, *Toxic Substances in Crop Plants*, The Royal Society of Chemistry, Cambridge, 1991.

34 C. S. Stewart, T. Acamovic, H. Gurung and A. S. Abdullah, in 'Manipulation of Rumen Microorganisms', Faculty of Agriculture, Alexandria University, Alexandria, Egypt, 1992, p. 150.

35 T. Acamovic, in *Developments and Ethical Considerations in Toxicology*, ed. M. I. Weitzner, The Royal Society of Chemistry, 1993, p. 129.

36 A. F. B. van der Poel, J. Huisman and H. S. Saini, *Recent Advances of Research in Antinutritional Factors in Legume Seeds*, Wageningen Pers, Wageningen, The Netherlands, 1993.

37 S. M. Colegate and P. R. Dorling, *Plant-Associated Toxins*, CAB International, Wallingford, 1994.

38 A. M. Craig, in *Ruminant Physiology: Digestion, Metabolism, Growth and Reproduction*, ed. W. v. Engelhardt, S. Leonhard-Marek, G. Breves and G. Giesecke, Ferdinand Enke, Stuttgart, 1995, p. 271.

39 R. W. A. W. Mulder, *Arch. Gerflugelkunde*, 1995, p. 55.

40 T. Acamovic, personal observation.

41 H. Karunajeewa, R. J. Hughes, M. W. McDonald and F. S. Shenstone, *Wlds. Poultry Sci. J.*, 1984, **40**, 52.

42 C. Juin, T. Acamovic, D. Younie and D. Yackiminie, *Occ. Publ. 17. Br. Soc. Anim. Prod.*, 1993, p. 83.

43 N. I. Krinsky, *Annu. Rev. Nutr.*, 1993, **13**, 561.

44 G. C. Smith, M. J. Aaronson, and J. N. Sofos, *Meat Focus Int.*, 1993, **2**, 463.

45 *Proceedings of 4th Annual Symposium of Biotechnology in the Feed Industry*, ed. T. P. Lyons, Alltech Technical Publications, Nicholasville, 1988.

46 *Proceedings of 7th Annual Symposium of Biotechnology in the Feed Industry*, ed. T. P. Lyons, Alltech Technical Publications, Nicholasville, 1991.

4 Effects in Animals

The effects of drugs in animals are usually concentration dependent and are also often animal species and site-of-action dependent;[2,4,16,48] thus the compounds may be present naturally in diets at low levels and produce no obvious adverse effects, either on the animal, gut microflora, meat, milk or eggs. The main purpose of the use of drugs in animal feeds is to improve the animal health and welfare and often to improve growth at minimum cost to the producer.[24] In the case of naturally produced compounds that may occur in the feedstuffs of animals, these compounds may impair animal health and performance as well as cause improvements.[37,47,49] The presence of such compounds in diets may be desirable or undesirable from the perspective of the animal and the farmer, but their presence in the respective products such as meat, milk and eggs frequently is prohibited, not desirable, and often restricted in quantity.[2,8,13,39,44,50-52] In the case of 'organic' produce the presence of drugs is strictly controlled, which can in some circumstances lead to poorer animal health and increased parasite burdens within the animal.[49]

Compounds in animal diets can exert their effects in the gastrointestinal tract or after absorption through the gut epithelial tissue into the bloodstream, or both.[48,53] On some occasions the administered compounds may be applied topically (Table 1).

A wide variety of animal species are subjected to the administration of drugs during their lifetime.[1,2,8,48] The various animal species can encounter drugs and other dietary additives by different routes and this is dependent on the environment in which they are kept. Intensively reared animals tend to have considerable consistency in the components of their diets and thus are much less likely to encounter the range of naturally produced compounds that extensively produced animals encounter. The desire for less expensive dietary constituents and increased efficiency of use has induced feed manufacturers and producers to add enzyme supplements to diets of most farmed animals to reduce the negative effects of indigestible dietary carbohydrates, refactory proteins and unavailable minerals such as phosphorus. This use of dietary additives to improve nutrient utilization and environmental consequences of feeding animals intensively has been the subject of intense research activity in the last five years.[11,48,54-63] The

[47] A. F. Erasmuson, B. G. Scahill and D. M. West, *J. Agric. Food Res.*, 1994, **42**, 2721.

[48] W. N. Ewing and D. J. A. Cole, *The Living Gut*, Context, Dungannon, 1994.

[49] G. C. Waghorn, G. Douglas, J. Niezen, W. McNabb and Y. Wang, *Proc. Soc. Nutr. Physiol.*, 1994, **3**, 30.

[50] T. J. Nicholls, in *Plant-Associated Toxins*, ed. S. M. Colegate and P. R. Dorling, CAB International, Wallingford, 1994, ch. 14, p. 71.

[51] A. A. Seawright, in *Plant-Associated Toxins*, ed. S. M. Colegate and P. R. Dorling, CAB International, Wallingford, 1994, ch. 15, p. 77.

[52] L. F. James, K. E. Panter, R. J. Molyneaux, B. L. Stegelmeier and D. J. Wagstaff, in *Plant-Associated Toxins*, ed. S. M. Colegate and P. R. Dorling, CAB International, Wallingford, UK, 1994, ch. 16, p. 83.

[53] J. W. Sissons, *J. Sci. Food Agric.*, 1989, **49**, 1.

[54] A. W. Jongbloed and P. A. Kemme, *Anim. Feed Sci. Technol.*, 1990, **28**, 233.

[55] G. Annison, *Anim. Feed Sci. Technol.*, 1992, **28**, 105.

[56] H. L. Classen and M. R. Bedford, in *Recent Advances in Animal Nutrition*, ed. W. Haresign and D. J. A. Cole, Butterworth-Heinemann, Oxford, 1991, p. 95.

Table 1 Examples of feed additives and drugs, their uses and site of action*†

Compound	Site	Purpose
Tropical		
Copper sulfate	Feet/hooves	Prevention of footrot
Zinc sulfate	Feet/hooves	Prevention of footrot
Organophosphorus compounds	Body	Insect and parasites
Diet		
Copper sulfate	Tissue/GIT‡	Prevention of Cu deficiency
Vitamins	Tissue/GIT	Prevention of vitamin deficiency
Monensin	Tissue/GIT	Coccidiostat and growth promoter
Enzymes	GIT	Improved utilization of feed
Yield promoters		
Somatotropins	Body tissue	Increased milk production
Stilboestrol	Reproductive organs	Improved reproductive capacity
Progestogens	Reproductive organs	Improved reproductive capacity
Zeranols	Body tissue	Improved body yield
Copper sulfate	GIT	Improved body yield

*Compiled from refs. 1, 2, 16 and 25.
†There may be constraints in the use in some countries.
‡GIT = gastrointestinal tract.

most successful of the supplementary enzymes are the β-glucanases, which effectively reduce the adverse effects of β-glucans in barley and rye-based diets, and the phytases which increase the availability of phytate-bound phosphorus in diets.[11] The latter enzyme has the effect of decreasing the amount of inorganic phosphorus, and other mineral ions, added to animal diets and thus the amount of phosphorus and other minerals excreted and concomitant reduction in pollution.[48,60,64–66] The supplementary dietary enzymes can act in the diet prior

[57] H. L. Classen, D. Balnave and M. R. Bedford, in *Recent Advances of Research in Antinutritional Factors in Legume Seeds*, ed. A. F. B. van der Poel, J. Huisman and H. S. Saini, Wageningen Pers, Wageningen, The Netherlands, 1993, p. 501.

[58] M. Choct and G. Annison, *Br. J. Nutr.*, 1992, **67**, 123.

[59] A. Chesson, *Anim. Feed Sci. Technol.*, 1993, **45**, 65.

[60] M. Dungelhoef, M. Rodehutscord, H. Spiekers and E. Pfeffer, *Anim. Feed Sci. Technol.*, 1994, **49**, 1.

[61] M. I. Ferraz de Oliveira and T. Acamovic, in *Proceedings of VII Symposium on Protein Metabolism and Nutrition, Evora, Portugal 1995*, 1996 in press.

[62] M. I. Ferraz de Oliveira and T. Acamovic, *Br. Poult. Sci.*, 1995, **35**, p. 841.

[63] M. B. Salawu, T. Acamovic, J. R. Scaife and W. Michie, *Br. Poult. Sci.*, 1995, **35**, p. 867.

[64] J. Venekamp, A. Tas and W. A. C. Somers, in *Proceedings of the 2nd European Symposium on Feed Enzymes, Noordwijkerhout, Netherlands*, 1995, p. 151.

[65] E. T. Kornegay, in *Proceedings of the 2nd European Symposium on Feed Enzymes, Noordwijkerhout, Netherlands*, 1995, p. 189.

[66] A. W. Jongbloed, P. A. Kemme, Z. Mroz and R. ten Bruggecante, in *Proceedings of the 2nd European Symposium on Feed Enzymes, Noordwijkerhout, Netherlands*, 1995, p. 198.

to ingestion and endogenously, in the gastrointestinal tract, thereby increasing efficiency and reducing pollution by reducing the waste nutrients voided in the excreta.[11,48] Enzyme supplementation of diets has been shown to improve efficiencies of feed and nutrient utilization in monogastric animals by up to 211%, but this is dependent on the nutrient and the type of diet used.[11,60–63,67]

The nature of the conditions of intensive production, however, can increase the risk of diseases and infections which can spread very rapidly and devastate large numbers of animals.[48,53] Thus it is common practice for producers of poultry to add coccidiostats to their diets and vaccines to their drinking water in order to prevent coccidiosis and other infectious diseases such as bronchitis and Newcastle disease. A similar problem exists for intensively reared fish, where it is necessary to add antibiotics to their diets. A problem with intensively reared fish is that their diet is added directly into the water in which they live; thus drugs and other additives in the diet are relatively easily dispersed into the local environment of fish farms, where they can increase bacterial resistance and also cause problems such as algal blooms.

Because the diets for intensively reared animals are relatively invariate in their composition, it is often necessary to add vitamins, minerals, antioxidants and pigmenters. The addition of pigmenters is particularly important in fish diets and in poultry diets, where pigmented tissue is preferred by some consumers. Such pigments can be man-made or obtained from natural products. Similarly, for cattle that are intensively produced and do not have access to grass, it may be necessary to add carotenoids to the diet to give appropriately pigmented tissue and creamy looking milk, which is preferred by some consumers. The addition of vitamins and some precursors to farmed animal diets is common practice around the world and, although in moderate quantities they have substantial beneficial effects on the health of animals and on quality of the products,[43] in recent years some concern has been expressed regarding excessive use of these in animal diets because of the risks of their deposition in body tissues which may then be consumed by humans. An example of such concern was the consumption of liver by pregnant women. It was found that some liver samples contained relatively high concentrations of vitamin A (retinol and its esters), which can be toxic to foetuses when ingested in large amounts.[3]

Farmed animals tend to be subjected to compounds that can be classified more easily as drugs. In some cases, animals may encounter compounds in nature that have considerable limitations in inclusion in the diets of farmed animals. An example is the inclusion of zearalenone (Figure 1) and its derivatives in animal diets; these have anabolic and oestrogenic properties[68] and are permitted to be used in some areas of the world but are prohibited from use in others, such as European Union countries.[8,44] Thus this compound and its reduced isomers (zearalenols, zeranols) improve growth rates in animals but can cause reproductive problems, especially in pigs. Ralgro, which is a commercially prepared derivative of zearalenone, is used as an ear implant in beef cattle as a growth promoter in

---

[67] M. Nasi, *Proceedings of 4th Annual Symposium of Biotechnology in the Feed Industry*, Alltech Technical Publications, Nicholasville, 1988, p. 199.

[68] R.J. Cole and R.H. Cox, *Handbook of Toxic Fungal Metabolites*, Academic Press, New York, 1981, p. 902.

some parts of the world. However, zearelenone is a mycotoxin that is produced by fungi of the genus *Fusarium* and which animals can encounter under extensive conditions as well as in contaminated concentrate feeds. Thus zeranols can be found naturally in animals at similar levels to animals that have been implanted with Ralgro.[13,47,68] The economic and legal consequences of the presence of such compounds in untreated animals could be considerable. There is significant opportunity for exposure to zearelenone in environments where it is not quantified and where animals have free access to feeds contaminated with *Fusarium*. As a consequence, increased weight gains may occur, but may be accompanied by reduced reproductive capacity.[69,70]

A natural plant amino acid that has received considerable attention over the last two decades is mimosine, which is found in plants of the Mimosoidaea family, particularly *Leucaena leucocephala*. It was found to cause depilation in mammals and was studied as a potential chemical defleecing agent.[71] Its mechanism of action is on hair follicles and has been reported in detail.[72,73] The hydrolysis product 3,4-DHP has potent goitrogenic and metal ion chelating properties which have been detailed in the literature.[74-76] *Leucaena* is a high-protein feedstuff used especially for browsing ruminant animals; thus animals grazing the material have often suffered the consequences of ingestion of the active principles, mimosine and 3,4-DHP (Figure 1). Some adverse effects encountered include poor performance, hair loss, reproductive problems and death, although chelation with metal ions appears to reduce some of the adverse effects.[76-78] Ruminant animals in some parts of the world did not suffer the problems of ingestion of mimosine and DHP due to the presence of microbial degradation within the rumen. Transfer of these mimosine- and DHP-degrading rumen microbes to susceptible animals has overcome some of these problems and there is a considerable research effort investigating other possibilities of utilizing either naturally occurring microbes or genetically manipulated microbes as degraders of xenobiotic compounds. The environmental consequences of providing animals with altered gut microbial populations are considerable and are discussed in detail later (Section 13 below). Other micro-organisms (sometimes known as probiotics) have been added to animal diets to improve health and growth but their effectiveness is variable.[48,53]

5 Environmental Effects

The environmental effects of using drugs and additives in animal diets are

[69] B. D. King, G. A. Bo, R. N. Kirkwood, C. L. Guenther, R. D. H. Cohen and R. J. Mapletoft, *Can. J. Anim. Sci.*, 1994, **74**, 73.
[70] B. D. King, G. A. Bo, C. Lulai, R. N. Kirkwood, R. D. H. Cohen and R. J. Mapletoft, *Can. J. Anim. Sci.*, 1995, **75**, 225.
[71] P. J. Reis, D. A. Tunks and D. A. Chapman, *Aust. J. Biol. Sci.*, 1975, **28**, 69.
[72] K. A. Ward and R. L. N. Harris, *Aust. J. Biol. Sci.*, 1976, **29**, 189.
[73] P. H. Li, *Life Sci.*, 1987, **41**, 1645.
[74] G. S. Christie, C. P. Lee and M. P. Hegarty, *Endocrinology*, 1979, **105**, 342.
[75] H. Stunzi, D. D. Perrin, T. Teitei and R. L. N. Harris, *Aust. J. Chem.*, 1979, **32**, 21.
[76] R. Puchala, T. Sahlu, J. J. Davis and S. P. Hart, *Anim. Feed Sci. Technol.*, 1995, **55**, 253.
[77] J. P. F. D'Mello and T. Acamovic, *Anim. Feed Sci. Technol.*, 1982, **7**, 247.
[78] T. Acamovic, PhD Thesis, University of Edinburgh, 1987.

considerable. In some cases, plant species are grown where they otherwise would not be grown. This can have beneficial effects on the environment where, in the case of the production of legumes, the soil can be stabilized and enriched in nitrogen. Because of the growth and utilization of forage from tree legumes such as *Leucaena*, erosion can be reduced. The provision of such alternative feedstuffs can reduce the overgrazing that can occur on grass-based pastures, especially when environmental conditions are not favourable for good grass production.

Dietary additives can affect the microbiota that are associated with the faeces of animals and degradation of the faeces may be impaired because of the influence of the excretory products on insects, microbes and fungi. The microbiota in the soil and waste material may be affected,[79] thus altering the fertility of the pasture and sustainability of other wildlife. These microbiota can be used as dietary ingredients for animals, so inhibition of their production would be an unsatisfactory consequence of dietary additives.[80]

Relatively high levels of copper in pig diets can improve nutritional performance due to the antimicrobial effects in the gastrointestinal tract.[2] However, if land is fertilized with dung from pigs and subsequently grazed by sheep, the sheep may suffer copper toxicity because of their increased susceptibility to copper compared with pigs. Similarly, pig diets would be unacceptable for sheep because of the high levels of copper therein.

In many societies the non-use or limited availability of drugs and dietary additives can mean that some populations around the world suffer the effects of malnutrition because of inadequate supply of protein. Businesses can fail, especially small businesses in the developing world, if the animals die because of lack of drugs. The administration of prophylactic drugs such as ivermectin, monensin, oxytetracyclines and many others, as well as dietary additives, can improve animal health and welfare, productivity and efficiency of productivity, thus reducing the incidence of pollution; a healthy, efficient animal utilizes dietary constituents more effectively, thus reducing the amount of excreted material for disposal.[48] This is particularly important in societies where intensive animal production is practised and imported feeds are used. On the other hand, in extensive systems it is necessary that dung of good quality, which has a minimal effect on the microbiota, is returned to the soil to maintain and enhance soil fertility.

Naturally occurring drugs are becoming more popular because of the perceived problems associated with the use of man-made drugs.[48] These naturally occurring compounds can have beneficial as well as adverse effects in animals.[19,35] Thus 'organically' reared ruminants are produced on mixed pastures fertilized by manure from animals which have not been subjected to man-made drugs and dietary additives. This can be problematical since such pastures can contain weeds and a higher proportion of legumes; thus these animals can incur a greater incidence of health problems than conventionally reared animals. In some cases, bloat can occur because the proteins in legumes stabilize foam formed in the rumen. Animals that graze some pastures containing higher proportions of legumes can encounter compounds and their metabolites (Figure 1) that can cause reproductive problems, especially in sheep that are

[79] R. A. Roncali in *Ivermectin and Abermectin*, ed. W. C. Campbell, Springer, New York, 1989, p. 173.
[80] I. A. Ibáñez, C. A. Herrara, L. A. Velásquez and P. Hebel, *Animal Feed Sci. Technol.*, 1993, **42**, 165.

Table 2 Examples of drugs and period of withdrawal from animals prior to use as food*†

Drug	Period of withdrawal	Animal species
Spectinomycin	5 days prior to slaughter	Cattle, sheep, pigs, poultry
Spectinomycin	2 days prior to milking	Cattle
Monensin	None (not used for lactating animals)	Cattle
Monensin	3 days prior to slaughter	Poultry
$CuSO_4$	None	Pigs
Stilboestrol	Not permitted for food animals	Cattle, pigs
Clenbuterol	28 days prior to slaughter	Calves
Clenbuterol	Not permitted for food animals	Horses
Lasalocid	7 days prior to slaughter	Poultry, game birds, rabbits
Ampicillin	28 days prior to slaughter	Horses, cattle, sheep, pigs
Chlorphenvinphos	14 days prior to slaughter	Sheep
Dichlorvos	4 days prior to slaughter	Fish
Oxolinic acid	400 degree days	Fish

*Compiled from refs. 1, 2, 16 and 25.
†Withdrawal periods are in days or degree days; *e.g.* 400 degree days is 20 days at 20 °C.

allowed to graze such mixed pastures.[12,35,37] Thus the lack of administration of appropriate drugs in the correct quantities can cause health and welfare as well as production problems in animals.

The administration of some drugs for prophylactic and enhanced production purposes requires a period of withdrawal prior to slaughter of the animals and suitable analytical methods to ensure that their products are acceptable as food (Table 2).[2,16,22,28,30]

6 The Gut Microbial Flora

Ingested plant metabolites, drugs and other compounds are processed both by the animal and by the microbial flora of the gut. The composition and activities of the gut microflora vary greatly from one animal species to another and have been very extensively reviewed.[48,53,85-87] In true ruminants (sheep, cattle and deer) and in functional ruminants, such as camels and llamas, a mixed population of bacteria,

[81] J. A. Timbrell, *Principles of Biochemical Toxicology*, Taylor and Francis, London, 2nd edn., 1992.
[82] *Developments and Ethical Considerations in Toxicology*, ed. M. I. Weitzner, The Royal Society of Chemistry, Cambridge, 1993.
[83] B. J. Blaauboer, W. C. Mennes and H. M. Wortelboer, in *Developments and Ethical Considerations in Toxicology*, ed. M. I. Weitzner, The Royal Society of Chemistry, Cambridge, 1993, p. 60.
[84] P. J. van Bladeren, J. Ploeman and B. van Ommen, in *Developments and Ethical Considerations in Toxicology*, ed. M. I. Weitzner, The Royal Society of Chemistry, Cambridge, 1993, p. 69.
[85] D. Savage, *Annu. Rev. Microbiol.*, 1977, **31**, 107.
[86] A. Lee, *Adv. Microbial Ecol.*, 1985, **8**, p. 115.
[87] T. Mitsuoka, in *Medical and Dental Aspects of Anaerobes*, ed. B. I. Duerden, W. G. Wade, J. S. Brazier, A. Eley, B. Wren and M. J. Hudson, Science Reviews, Northwood, 1995, p. 87.

fungi, and protozoa is present; all of the fungi, many of the protozoa and around 6–8 (out of over 200) species of the bacteria are cellulolytic and therefore contribute digestive enzymes not produced by the animals.[88–90] The horse caecum and colon also contain protozoa, fungi and bacteria, although the species composition of the protozoal populations differs from that found in ruminants.[91–93] Like man, pigs have a predominantly bacterial flora, and fungi or protozoa are not found, though these organisms might be expected to establish in the hind gut of older pigs.[94] Gut protozoa are also found in the hind gut of the gorilla, hippopotamus and rhinoceros; in addition, they are found in certain insects such as termites.[95] The specific contributions of protozoa and fungi to the metabolism of xenobiotics are largely unexplored. However, the presence of protozoa has a substantial impact on the nutritional ecology of ruminants by virtue of their cellulose-degrading capacity.[96,97] The ciliate protozoa also predate bacteria and fungal zoospores. Protozoa are sequestered in the rumen, and this results in the retention of microbial nitrogen in the rumen, reducing the flow of nitrogen down the digestive tract.[98] When ruminants are fed diets low in dietary nitrogen, defaunation (elimination of rumen protozoa) may improve the flow to the small intestine.[99] The anaerobic rumen fungi produce some of the most active cellulases known to science; however, the contribution of these organisms to the rumen fermentation[100] remains unclear, as the cellulases of these fungi are inhibited by soluble proteins produced by some rumen bacteria of the genus *Ruminococcus*.[101,102]

7 Gastrointestinal Microorganisms and Feeding Strategy

Animals and their associated gut micro-organisms have co-evolved and it is therefore possible to implicate micro-organisms, directly or indirectly, in both

[88] R. E. Hungate, *The Rumen and its Microbes*, Academic Press, New York, 1966.

[89] R. E. Hungate, *The Rumen Microbial Ecosystem*, ed. P. N. Hobson, Elsevier Applied Science, London, 1988, pp. 1–19.

[90] K. Ogimoto and S. Imai, *Atlas of Rumen Microbiology*, Japan Scientific Societies Press, Tokyo, 1981.

[91] C. G. Orpin, *J. Gen. Microbiol*, 1975, **91**, 249.

[92] A. Breton, M. Dusser, B. Gaillard-Martinie, B. Guillot, N. Millet and G. Prensier, *FEMS Microbiol. Lett.*, 1993, **82**, 1.

[93] K. Ozeki, S. Imai and M. Katsuno, *Tohoku. J. Agr. Res.*, 1973, **24**, 86.

[94] C. S. Stewart, in *Ecology and Physiology of Gastrointestinal Microorganisms*, eds. R. I. Mackie and B. A. White, Chapman and Hall, New York, 1996, in press.

[95] R. H. McBee, in *Microbial Ecology of the Gut*, ed. R. T. J. Clarke and T. Bauchop, Academic Press, London, 1977, pp. 185–222.

[96] R. A. Prins, in *Nutrition and Drug Interrelations*, ed. J. N. Hathcock and J. Coon, Academic Press, New York, 1978, pp. 189–251.

[97] A. G. Williams and G. S. Coleman, in *The Rumen Microbial Ecosystem*, ed. P. N. Hobson, Elsevier Applied Science, London, 1988, p. 77.

[98] D. M. Veira, *J. Anim. Sci.*, 1986, **63**, 1547.

[99] C. J. Van Nevel and D. I. Demeyer, in *The Rumen Microbial Ecosystem*, ed. P. N. Hobson, Elsevier Applied Science, London, 1988, pp. 387–443.

[100] C. S. Stewart, M. Fevre and R. A. Prins, in *Ruminant Physiology; Digestion, Metabolism, Growth and Reproduction*, ed. W. von Englehardt, Ferdinand Enke, Stuttgart, 1985, pp. 249–268.

[101] C. S. Stewart, S. H. Duncan, R. J. Richardson, C. Backwell and R. Begbie, *FEMS Microbiol. Lett.*, 1992, **97**, 83.

[102] A. Bernalier, G. Fonty, F. Bonnemay and P. H. Gouet, *J. Gen. Microbiol.*, 1993, **139**, 873.

vertebrate and invertebrate evolution and feeding behaviour, which have obvious effects on the environment.[97] In particular, the localization of gut micro-organisms in different parts of the digestive tract can be shown to be related to different feeding strategies.[103] Feeding strategies in turn influence many aspects of anatomy, physiology and behaviour, most of which are beyond the scope of this article. Most obviously, the dentition is specialized to allow optimal physical processing of the diet, and the chewing of regurgitated digesta from the rumen (rumination) speeds the digestive process. The trunk of elephants and the extended neck of giraffes are means of food collection; the evolution of the equids (horses, zebras) shows a particularly clear pattern of development of the limbs, allowing rapid movement, both in search of food and also as a means of escape from carnivorous predators, which in turn include animals capable of movement at very high speeds.[104] The total numbers of bacteria in the gut, in which population densities may reach 10^{11} cm^{-3}, is usually greater than the number of cells in the host animal's body. The effect of the presence of this large microbial flora is influenced by the location of the micro-organisms in relation to the digestive functions of the host. Thus the crop of chickens and the rumen of ruminants and camelids is pre-peptic, whereas in man, horses and pigs the microbial population occupies a post-peptic location, the large intestine (Table 3). The location of the first gut compartment in which the digesta is exposed to microbial digestion is of particular significance, because these micro-organisms ferment readily available carbohydrates and incorporate dietary nitrogen into microbial cells, thus affecting the pattern of carbohydrate and nitrogen metabolism. When microbial protein is synthesized in foregut compartments, it can provide digestible protein to the animal. Microbial proteins synthesized in the hind gut are not available to the animal. In all animals, some fermentation inevitably occurs in the hindgut, though it is minimal in animals with an unsacculated colon. The panda, for example, is a member of the order Carnivora which eats vegetation (bamboo[105]). Cellulose fermentation in the gut is relatively poor, however, and these animals seem to depend upon extracting the readily available substrates, processing a large amount of plant material in order to satisfy their nutrient requirements. In many rodents and in rabbits, coprophagy provides a means of recovery of nitrogen from the faeces.

8 Microbial Fermentation, CO_2 and CH_4 Formation

Carbon dioxide and methane are among the most important greenhouse gases. There is particular interest in the contribution of animal production to the greenhouse effect and it has been calculated that ruminants may account for around 15–20% (estimates range from around 7% to around 27%) of the annual global production of methane. By comparison, pigs, man, and other monogastric animals produce relatively little methane, but termites and other insects are substantial producers[106] (Figure 2). Gut micro-organisms ferment polysaccharides

[103] P. J. Van Soest, *Nutritional Ecology of the Ruminant*, Cornell University Press, Ithaca, 2nd edn., 1994.

[104] J. Z. Young, *The Life of Vertebrates*, Oxford University Press, Oxford, 2nd edn., 1962.

[105] E. S. Dierenfield, H. F. Hintz, J. B. Robertson, P. J. Van Soest and O. T. Oftedal, *J. Nutr.*, 1982, **112**, 636.

[106] P. J. Crutzen, I. Aselmen and W. Seiter, *Tellus*, 1986, **388**, 271.

Table 3 Location of primary microbial populations in the gut*

Group	Species	Feeding strategy
Pregastric fermenters		
Ruminants (functional)	Cattle, sheep	Grazing herbivore
	Antelope, deer, camel, llama	Selective herbivore
Non-ruminants	Colobine monkey	Selective herbivore
	Hamster, vole	Selective herbivore
Birds	Chicken	Omnivore
	Hoatzin	Folivore
Hindgut fermenters		
Caecal digesters	Capybara	Grazer
	Rabbit	Selective herbivore
	Rats and mice	Omnivores
Colonic digesters		
Sacculated	Elephant, equines	Grazers
	New world monkeys	Folivores
	Pig, human	Omnivores
Unsacculated	Dog, cat	Herbivore
	Panda	Herbivore

*Modified from Van Soest, 1994.[103]

to form volatile fatty acids, CO_2 and hydrogen. In ruminants and many other animals, methane may be formed from the reduction of CO_2 or from formate by a specialized population of methanogenic bacteria. The stoichiometry of product fermentation is similar in ruminants and many monogastric animals, including man:[107]

Rumen contents
$$57.5\,C_6H_{12}O_6 \rightarrow 65\,CH_3CO_2H + 20\,CH_3CH_2CO_2H$$
$$+ 15\,CH_3(CH_2)_2CO_2H + 35\,CH_4 + 60\,CO_2$$

Human faeces (non-methanogenic)
$$54\,C_6H_{12}O_6 \rightarrow 55\,CH_3CO_2H + 22\,CH_3(CH_2)_2CO_2H + 46\,H_2 + 57\,CO_2$$

Human faeces (methanogenic)
$$57.5\,C_6H_{12}O_6 \rightarrow 56\,CH_3CO_2H + 21\,CH_3CH_2CO_2H$$
$$+ 19\,CH_3(CH_2)_2CO_2H + 32.25\,CH_4 + 61.75\,CO_2$$

Hackstein *et al.*[108] have suggested that there may be a phylogenetic basis for the occurrence of significant methane formation in the gut of animals. It was argued that this reflected the presence of a 'methanogen receptor' in the gut of animals which support large populations of these bacteria, since the production of

[107] M. J. Wolin and T. L. Miller, in *Human Intestinal Microflora in Health and Disease*, ed. D. J. Hentges, Academic Press, London, 1983, pp. 147–165.
[108] J. H. P. Hackstein, T. A. van Allen, H. Op den Camp, A. Smits and E. Mariman, *Dtsch. Tieraertzl. Wschr.*, 1995, **102**, 152.

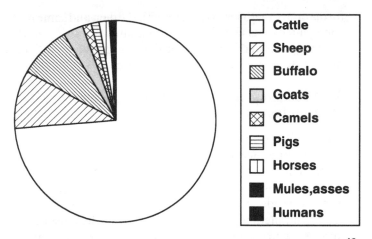

ure 2 Estimated methane
·duction by domesticated
animals and humans.
(From data of Crutzen
et al., 1986)

(approx. total 73.7 x 10^{12} g/year)

methane is roughly proportional to the size of the methanogenic population. Whether or not large populations of methanogens are maintained may also depend upon the presence of other bacteria such as acetogens or reducers of nitrate or sulfate, which compete with methanogens for reducing equivalents for CO_2 reduction.[109,110] The presence of these populations may be controlled by the host animal. For example, sulfate reducers utilize sulfate released from sulfated polysaccharides in host secretions; the composition and production of these secretions is presumably under genetic control and could be the basis of the link between the phylogeny of the animals and the presence of vigorous methanogenesis. Alternatively, methanogens may rely on other bacteria for the supply of some nutrients for growth. The production of methane by ruminants represents a waste of energy to the animal. The disposal of protons and carbon as CH_4 reduces the production of reduced products such as propionate. Methanogenesis can be reduced by the use of dietary additives, including chloral hydrate which forms the methane analogue chloroform in the rumen, and a number of antibiotics. Antibiotics used for this purpose vary in their mode of action. For example, the glycopeptide avoparcin inhibits the synthesis of peptidoglycan, whereas the ionophores monensin, lasalocid and tetronasin permeabilize bacteria to cations, including protons, thus dissipating trans-membrane proton gradients. In general, Gram positive bacteria are more susceptible to these compounds than are Gram negative bacteria[111-113] and the basis of the inhibition of methanogenesis is simply that many (but not all) of the rumen bacteria that produce H_2 have Gram positive cell walls, including *Ruminococcus, Butyrivibrio* and *Lachnospira*. The

[109] J. G. Morris, in *Anaerobic Bacteria in Habitats Other than Man*, ed. E. M. Barnes and G. C. Mead, Blackwell Scientific Publications, Oxford, 1986, pp. 1–21.
[110] J. Dolfing, in *Biology of Anaerobic Microorganisms*, ed. A. J. B. Zehnder, Wiley, New York, 1988, pp. 417–468.
[111] M. Chen and W. J. Wolin, *Appl. Environ. Microbiol.*, 1979, **38**, 72.
[112] C. S. Stewart, M. Crossley and S. H. Garrow, *Eur. J. Appl. Microbiol. Biotechnol.*, 1983, **7**, 292.
[113] C. J. Newbold, R. J. Wallace, N. S. Watt and A. J. Richardson, *Appl. Environ. Microbiol.*, 1988, **54**, 544.

formation of methane also varies with the diet, and some recent work suggests that some of the diets used for intensive animal production, which often incorporate starchy grains and other concentrates, support less methane production than the fibrous diets of extensively produced animals.[114]

9 Methanogenesis from Acetate, Residence Time and Size of Ruminants

The methanogens found in the gut include species that are able to produce methane from acetate.[115] However, the metabolism of acetate in the rumen to form methane would represent a marked penalty for the animal, which is able to use acetate for biosynthetic purposes. Methanogenesis from acetate is very slow and the residence time of digesta in the gut is normally too short to allow significant methanogenesis from acetate. It has been argued that this reaction imposes an upper limit on the maximum size of ruminants; ruminants as large as elephants (which are monogastric), or the large herbivorous dinosaurs, would retain their digesta so long that acetate could be converted to methane by methanogens like *Methanosarcina*, thus depriving the ruminant of an important nutrient.[103] Acetate in faeces is presumably oxidized by aerobic or facultative soil microorganisms; where animal waste is discharged into natural or engineered anoxic environments, acetate and other substrates support methanogenesis.[116]

10 Methane and Ammonia Oxidation

The production of reduced products like methane and ammonia by the gut microbial flora has important environmental consequences, as such compounds contribute to the chemical and biological oxygen demand. The detection of dissolved oxygen in the gut of piglets[117] led to tests to show whether methane and other reduced products could be oxidized in the pig gut. The production of ^{13}C-labelled CO_2 from ^{13}C-labelled methane has been demonstrated;[118] however, it is calculated that at most only a very small proportion of methane produced is likely to be oxidized using O_2 as electron acceptor. Methane may also be oxidized anaerobically,[119,120] but only an extremely small amount of methane is likely to be oxidized in this way in the gut.

Tests on pig gut contents using molecular probes to detect the presence of (aerobic) ammonia oxidizers proved negative.[121] Recently, the anaerobic oxidation of ammonia coupled to nitrate reduction has been demonstrated in

[114] M. Kirschgassner, W. Windisch and H. L. Müller, in *Ruminant Physiology: Digestion, Metabolism, Growth and Reproduction*, ed. W. U. Engelhardt, S. Leonhardt-Marek, G. Breves and D. Gieseke, Ferdinand Enke, Stuttgart, 1995, p. 333.

[115] D. Archer and J. E. Harris, in *Anaerobic Bacteria in Habitats Other than Man*, ed. E. M. Barnes and G. C. Mead, Blackwell Scientific Publications, Oxford, 1986, pp. 185–223.

[116] G. B. Kasali, S. B. Shibani and E. Senior, *Lett. Appl. Microbiol.*, 1989, **8**, 37.

[117] K. Hillman, A. Whyte and C. S. Stewart, *Lett. Appl. Microbiol.*, 1993, **16**, 299.

[118] K. Hillman, H. van Wyk, E. Milne, C. S. Stewart, and M. F. Fuller, in *Symposium on Global Methane Flux, Abstracts*, Society for General Microbiology, 1995, p. 24.

[119] A. J. B. Zehnder and T. D. Brock, *J. Bacteriol.*, 1979, **137**, 420.

[120] A. J. B. Zehnder and T. D. Brock, *Appl. Environ. Microbiol.*, 1980, **39**, 194.

[121] S. Duncan, K. Hillman, A. Prosser and C. S. Stewart, unpublished data.

Table 4 Biotransformation reactions by the gastrointestinal microflora

Reaction	Examples
Hydrolysis	Polysaccharides, glycosides, proteins, lipids, esters
Reduction	C = C, N = N, aldehydes, ketones, alcohols
Degradation	Decarboxylation, deamination, dehalogenation, dehydroxylation, ring fission, demethoxylation, deacetylation
Conjugation	Esterification

anaerobic digesters.[122] Whether this process could occur in the gut has not been tested.

11 Microbial Transformations and Interactions with the Xenobiotic Metabolizing Enzyme (XME) System

Some of the principal reactions involved in the microbial transformation of nutrients by bacteria in the gut are summarized in Table 4, compiled from other reviews[123,124] which include descriptions and examples of the processes involved. These processes provide microorganisms with energy sources, C-skeletons and other nutrients for growth. The host animals benefit from the formation of volatile fatty acids, which can provide a source of energy and C-skeletons for biosynthesis, the provision of B vitamins, physical comminution and processing of the digesta which in turn influence digesta passage and gut clearance, and the detoxification of metabolites which would otherwise cause adverse effects. The xenobiotic metabolizing enzyme (XME) system is mainly present in the liver, but is also found in other organs. An excellent text on the principles of xenobiotic metabolism and reaction of drugs in mammalian systems is available,[81] while another publication explores a number of aspects of xenobiotics in mammalian diets and their environmental effects.[82] Some of the elements of the XME system and the types of reactions of the gut bacteria that may interact with this system are summarized schematically in Figure 3. The XME system is responsible for the detoxification of drugs and other metabolites, enhancing their hydrophilicity and facilitating their excretion.[4,81,96] Oxidations of xenobiotics are normally catalysed by cytochrome P450 oxidases, then conjugated with polar radicals such as glutathione or glucuronic acid, prior to excretion.[81,83,84,125] However, the action of cytochrome oxidase may result in the production of reactive metabolites which bind to nucleophilic macromolecules (protein, DNA, RNA). The conjugated derivative may occasionally be more reactive and toxic than the starting compound, as is the case with morphine-6-gluconuride, the conjugate derivative of morphine.[81,126] Gastrointestinal bacteria may metabolize xenobiotic compounds

122 A. Van de Graaf, A. Mulder, P. de Bruijn, M. J. S. Jetten, L. A. Robertson and J. G. Kuenen, *Appl. Environ. Microbiol.*, 1995, **61**, 1246.

123 R. R. Scheline, *Pharmacol. Rev.*, 1973, **25**, 451.

124 R. A. Prins, in *Microbial Ecology of the Gut*, ed. R. T. J. Clarke and T. Bauchop, Academic Press, London, 1977, pp. 73–184.

125 I. M. C. M. Reitjens, N. H. P. Cnubben, P. A. de Jager, M. G. Boersma and J. Vervoort, in *Developments and Ethical Considerations in Toxicology*, ed. M. I. Weitzner, The Royal Society of Chemistry, Cambridge, 1993, p. 94.

Figure 3 Some interactions between gastrointestinal bacteria and the xenobiotic metabolizing enzyme system of the host. Solid lines, host XME system reactions; dotted lines, classes of reactions mediated by gastrointestinal bacteria. Specific examples of reaction classes 1–5 are cited in the text. (Modified from Rowland and Tanaka[127])

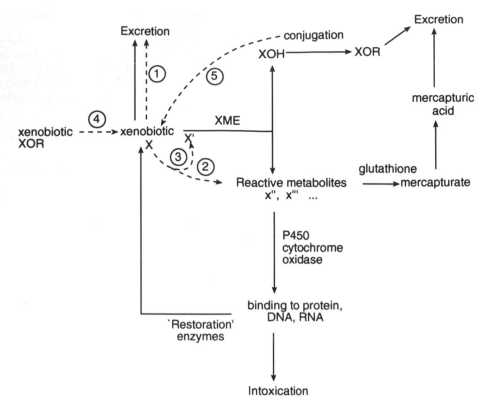

at various stages during their metabolism by the XME system. In Figure 3, five different classes of bacterial reactions have been defined, depending upon the point at which bacteria interact with the XME system of the host. This is an arbitrary classification designed to illustrate the range of reactions which may be involved.

Class 1 reaction—metabolism of a xenobiotic prior to excretion. Methylmercury is demethylated by faecal microorganisms, prior to excretion as inorganic mercury. *Class 2 reaction*—direct microbial formation of a reactive metabolite. In many cases, gut bacteria may form a product which is harmful, such as cyclohexylamine which is formed by the desulfination of cyclamate. *Class 3 reaction*—the formation of a xenobiotic derivative. Monoazo dyes and some nitro compounds may be transformed via bacterial azo- and nitro-reductases to aromatic amines which are metabolized by the XME system to carcinogens. *Class 4 reaction*—direct bacterial hydrolysis of ingested glycosides prior to metabolism by the XME system. Examples include the formation of the carcinogen methylazoxymethanol from cycasin, a glycoside in cycad nuts, and hydrolysis of the cyanogenic glycoside amygdalin. *Class 5 reaction*—the hydrolysis of XME products. This is exemplified by the carcinogen *N*-hydroxyfluorenylacetamide, which is detoxified by the XME system and excreted

[126] N. Roland, L. Nugon-Baudon and S. Rabot, in *Intestinal Flora, Immunity Nutrition and Health*, ed. A. P. Simopoulos, T. Corring and A. Rerat, World Reviews of Nutrition and Dietetics, Karger, Basel, 1993, vol. 74, pp. 123–148.

in the bile as a glucurono conjugate. The hydrolysis of this compound by bacteria in the large intestine releases the carcinogenic aglycone. Colonic bacterial transformation reactions may thus release reactive metabolites into the colon which may act as carcinogens or pro-carcinogens. The inhibition of reactions involved in the release of carcinogens is a desirable property of protective factors in the diet. Foods containing lactic acid bacteria have proved to have useful properties in this respect. In general, the addition of lactic acid bacteria ('probiotics') to the diet tends to decrease colonic glucuronidase activity. For example, the introduction of *Bifidobacterium breve* was shown to substantially lower the large intestinal glucuronidase activity.[127] Among farm animals, there is particular interest in the use of lactic acid bacteria and other 'probiotics' in pigs, poultry, and calves.[128]

12   Toxic Constituents of Plants

Toxic secondary metabolites of plants, sometimes referred to as allelochemicals,[129] reduce their susceptibility to competition from other plants and to predation by insects, birds, and other animals. Some of these protective compounds can be metabolized by gut bacteria. These bacteria are not always very widely distributed, and differences can arise between the ability of animals in different geographical regions to utilize plants containing some such metabolites. For example, the amino acid mimosine is present in the leaves and seeds of the legume *Leucaena leucocephala*. Australian ruminants are susceptible to mimosine and its degradation products 2,3-DHP and 3,4-DHP, and ingestion of *Leucaena* is accompanied by the development of toxic symptoms. However, the inoculation of Australian ruminants with rumen bacteria (isolated from animals in Hawaii and Indonesia) capable of degrading DHP has allowed the inoculated animals to consume *Leucaena* without ill-effects. Mimosine and its toxic derivatives can be degraded by several types of bacteria, including *Synergistes jonesii*, a new genus and species isolated by Allison and his colleagues,[130] at least one *Clostridium* species and other, so far unidentified, strains.[131,132] The detoxification of such compounds by introduced bacteria demonstrated that the rumen flora shows at least some geographical variation, and led to increased interest in the nature of plant toxic factors and their microbial detoxification. In particular, experiments are in progress to develop genetically engineered strains capable of the anaerobic degradation of the toxin fluoroacetate, a component of some *Gastrolobium* and *Oxylobium* species, and bacteria capable of transforming other potentially toxic plant components are being sought.[133] In fact, anaerobic dehalogenation

[127] I. R. Rowland and R. Tanaka, *J. Appl. Bacteriol.*, 1993, **74**, 667.
[128] F. J. Maxwell and C. S. Stewart, in *The Neonatal Pig*, ed. M. Varley, CAB, Wallingford, 1995, pp. 155–186.
[129] M. F. Balandrin, J. A. Klocke, E. S. Wurtele and W. H. Bollinger, *Science (Washington, DC)*, 1985, **228**, 1154.
[130] M. J. Allison, W. R. Mayberry, C. S. McSweeney and D. A. Stahl, *Syst. Appl. Microbiol.*, 1992, **15**, 522.
[131] M. G. Dominguez-Bello and C. S. Stewart, *FEMS Microb. Ecol.*, 1990, **73**, 283.
[132] M. G. Dominguez-Bello and C. S. Stewart, *Syst. Appl. Microbiol.*, 1991, **14**, 67.
[133] K. Gregg and H. Sharpe, in *Physiological Aspects of Digestion and Metabolism in Ruminants*, ed. T. Tsuda, Y. Sasaki and R. Kawashima, Academic Press, San Diego, 1991, pp. 719–735.

reactions are not unusual and it is rather surprising that genetic manipulation was employed for this purpose.

Other plant components may inhibit the fermentation of plant material in the rumen by a variety of different mechanisms. Hydroxylated compounds, in particular, may bind to proteins; the binding of such compounds may inhibit enzyme activity. The inhibition of rumen cellulase by a water-soluble extract from the forage *Sericea* has been described.[134] Polyphenolic compounds were implicated when it was found that the inhibitory effects on hydrolysis of carboxymethylcellulase were suppressed by treatment of the extracts with polyphenol oxidases. Phenolics, tannins, and leucoanthocyanins were assumed to be responsible for the inhibition of pectinases by extracts from grapes and other plants.[135] Of over 200 samples of fresh leaves, flowers, fruit and seeds tested, significant inhibitory activity towards cellulase was detected in about 20%. The leaves of bayberry (*Myrica pensylvanica*), muscadine grape (*Vitis rotundifolia*), *Vaccinium nitidum, Comptonia peregrina, Lespedeza cuneata, Diospyros virginiana* and other plants were found to contain cellulase inhibitors. The inhibitor from bayberry was shown to bind readily to proteins and to act as a non-competitive enzyme inhibitor. Several natural tannins and the lower molecular mass phenols haematoxylin, 3,5,4'-trihydroxystilbene and its glucoside, pyrogallol, and tannic acid were also shown to inhibit Trichoderma cellulase.

In many areas in which grasses grow poorly, such as arid, semi-arid and cold mountainous areas, leaves of trees and shrubs are fed to ruminants. Tree leaves often contain both condensed and hydrolysable tannins.[136,137] These tannins may depress intake and ruminal fermentation, reduce wool growth, decrease nutrient availability (especially sulfur), and cause toxicity. Tannins precipitate proteins and inhibit many enzyme reactions non-specifically. Microbial enzymes inhibited by tannins from *Quercus incana* include urease, carboxymethyl cellulase, glutamate dehydrogenase and alanine aminotransferases. Curiously, the activity of glutamate ammonia ligase in the rumen was increased in the presence of *Quercus* tannins. Tannins from *Robinia* were shown to inhibit β-glucosidase activity, and tannins from *Ziziphus* inhibited casein hydrolysis. The plant phenolics *p*-coumaric and ferulic acid inhibit growth and cellulolysis by rumen bacteria in pure culture.[139] Coumarin (1,2-benzopyrone) inhibits glucose fermentation by rumen anaerobic fungi.[138] Although these compounds were tested at concentrations greater than would be found in the bulk phase of rumen contents, their local concentration might be high in the microenvironment of the degrading plant cell wall surface.[139] Apart from the possibility of introducing rumen micro-organisms capable of degrading plant toxins, identifying the toxins present in different plants is useful to plant breeders. Ruminant diets may also be supplemented with chemicals, such as poly(vinylpyrrolidine) and poly(ethylene glycol) that form complexes with phenolics, reducing their toxic effects.

[134] W. W. G. Smart, T. A. Bell, N. W. Stanley and W. A. Cope, *J. Dairy Sci.*, 1961, **44**, 1945.
[135] M. Mandels and E. T. Reese, in *Advances in Enzymic Hydrolysis of Cellulose and Related Materials*, ed. E. T. Reese, Pergamon, Oxford, 1963, pp. 115–157.
[136] R. Kumar and S. Vaithiyanathan, *Anim. Feed Sci. Technol.*, 1990, **30**, 21 038.
[137] H. P. S Makkar and K. Becker, *J. Agric. Food Chem.*, 1994, **42**, 731.

13 Antibiotics and Antibiotic Resistance

Drugs and other xenobiotics in the diet of animals have a variety of physiological effects, some of which have been outlined above. Such compounds may also affect the genetic composition of the organism. The ability of pharmaceuticals and other chemicals to cause genetic mutations may be screened by testing the frequency of formation of mutants by the gut bacterium *Salmonella*, the basis of the Ames test.[140] Although all organisms show genetic plasticity, the short generation time of gut bacteria (around 20 min in the most rapidly growing species) and the opportunity for the transfer of genes between micro-organisms by conjugation, transformation and transduction presents opportunities for the spread of genetic traits. Genes which may be transferred between micro-organisms include those encoding the ability to utilize certain compounds such as toluene, and those encoding the resistance to antibiotics.[141]

The use of antimicrobial drugs in animal feeds led to the emergence of drug-resistant strains of pathogenic gastrointestinal bacteria.[142] As a result of the desire to limit the spread of genes encoding resistance to therapeutic drugs, the feeding of tetracycline to pigs was discontinued in the UK in 1971. However, it was found subsequently that there was no significant decrease in the incidence of pigs harbouring tetracycline resistant *Escherichia coli* in their faeces by 1975.[143] In a study in the USA,[144] it was found that the withdrawal of antibiotics in pig feeds for over 10 years led to a reduction in the incidence of tetracycline-resistant bacteria of around 50%, a smaller fall than had been anticipated, and one that would probably have little practical impact on the possibility of the further spread of resistance factors. Resistance to antibiotics may be very widespread among gut bacterial populations. Thus chlortetracycline resistance was detected in many genera of bacteria isolated from the faeces of pigs fed this and other antibiotics, including strains of *Streptococcus, Bacteroides, Lactobacillus, Bifidobacterium, Eubacterium, Propionobacterium, Desulphomonas, Leptotrichia* and *Fusobacterium*,[145] although the contribution of mobilizable genetic elements to this scenario was not established.

The level of antibiotic resistance of the gut flora of pigs has been found to be influenced by factors other than the inclusion of antibiotics in the diet:[146] such factors include the herd environment, history and the opportunity for cross-contamination.

Cross-contamination between animal species may be a factor in the spread of resistance. Flint and Stewart[147] showed that a rumen strain (46/5) of the Gram

[138] E. Cansunar, A. J. Richardson, G. Wallace and C. S. Stewart, *FEMS Microbiol. Lett.*, 1990, **70**, 157.

[139] A. Chesson, C. S. Stewart and R. J. Wallace, *Appl. Environ. Microbiol.*, 1982, **44**, 597.

[140] B. N. Ames, P. E. Hartman and J. Jacob, *J. Mol. Biol.*, 1963, **7**, 23.

[141] S. J. Burt and C. R. Woods, *J. Gen. Microbiol.*, 1976, **93**, 405.

[142] H. W. Smith, *Nature (London)*, 1968, **218**, 728.

[143] H. W. Smith, *Nature (London)*, 1975, **258**, 628.

[144] B. E. Langlois, K. A. Dawson, T. S. Stahly and C. L. Cromwell, *Appl. Environ. Microbiol.*, 1983, **46**, 1433.

[145] B. Welch and C. W. Forsberg, *Can. J. Microbiol.*, 1979, **25**, 789.

[146] K. A. Dawson, B. E. Langlois, T. S. Stahly and G. L. Cromwell, *J. Anim. Sci.*, 1984, **58**, 123.

[147] H. J. Flint and C. S. Stewart, *Appl. Microbiol. Biotechnol.*, 1987, **26**, 450.

negative bacterium *Mitsuokella* (*Bacteroides*) *multiacidus* isolated in Scotland, and a strain (P208-58) isolated from the faeces of pigs in Japan, were resistant to tetracycline. Both strains carried single, small plasmids of around 12 kbp: whether the resistance genes were borne on the plasmid or the chromosome was not established. Subsequently, it was shown that the plasmid in strain 46/5 showed high homology with that in strain P208-58.[148] Bacteria like *M. multiacidus*, which has been isolated from the faeces of humans and pigs and also from the rumen, may provide a route for the spread of antibiotic resistance determinants between animal species *via* the gastrointestinal microbial populations. Many other bacterial species occur in a wide range of animals, perhaps none more so than *Escherichia coli*, some strains of which are pathogenic, whilst other strains are harmless commensals.[149] Antibiotic resistant strains of *E. coli* can be readily isolated from farm animals, including ruminants. Strains of *E. coli* isolated from ruminants were found to possess transmissible antibiotic resistance to tetracycline, ampicillin and streptomycin.[150] The transmissible multiple resistance was associated with plasmids of approx. 80 kbp. Non-transmissible resistance to tetracycline and streptomycin was also detected. Rumen strains of *E. coli* were shown to be capable of exchanging a plasmid encoding tetracycline and ampicillin resistance under rumen-like conditions (*i.e.* anaerobically, in the presence of rumen fluid) *in vitro*.[151] Tetracycline resistant strains of other rumen bacteria, including *Megasphaera elsdenii*, *Selenomonas ruminantium*, and *Butyrivibrio fibrisolvens*, were isolated.[152] Some of the strains of *M. elsdenii* and of *S. ruminantium* possessed plasmids, though whether the tetracycline resistance genes were plasmid-encoded was not established.

14 Conclusions

The continued and increasing requirement of humans for meat and animal products of high quality in sufficient quantities at low prices necessitates the provision of animals with adequate amounts of feedstuffs. The feedstuffs which they access vary widely throughout the world and are dependent on the environment and on the animal species considered. Extensively reared animals with low inputs can encounter a much wider variety of feedstuffs than those animals which are intensively reared. As a consequence, they may encounter dietary ingredients that have adverse or beneficial effects on their health and growth and on the microbiota that inhabit their gastrointestinal tract. Their contact with man-made dietary additives and drugs will thus be limited and, although contamination of their products by such compounds will be limited, they may encounter similar or even identical compounds which are present in their diet as a consequence of being a secondary metabolite within the feedstuff or

[148] A. Daniel and H. J. Flint, unpublished observation.
[149] I. Orskov and F. Orskov, in *Molecular Pathogenesis of Gastrointestinal Infections*, ed. T. Wadstrom, P. H. Makela, A.-M. Svennerholm and H. Wolf-Watz, Plenum, New York, 1991, pp. 49–53.
[150] H. J. Flint, S. H. Duncan and C. S. Stewart, *Lett. Appl. Microbiol.*, 1987, **5**, 47.
[151] K. P. Scott and H. J. Flint, *J. Appl. Bacteriol.*, 1995, **78**, 189.
[152] H. J. Flint, S. H. Duncan, J. Bisset and C. S. Stewart, *Lett. Appl. Microbiol.*, 1988, **6**, 113.

on fungi that were produced by spoilage fungi growing on the feedstuff. Such compounds may exert effects within the gastrointestinal tract on the tissues or enzymes or on the microflora that exists therein. They may also be absorbed and exert their effect systemically within the animal. In general, the waste products from extensively reared animals are environmentally beneficial, with the exception of the relatively large amounts of methane produced per animal. On the other hand, intensively reared animals produce large amounts of waste material in relatively small areas and thus can be environmentally disadvantageous. Drugs and dietary additives that help to reduce pollution from such sources are advantageous; however, their use may be considered less acceptable because of deposition in animal products and the transfer of antibiotic resistance genes to micro-organisms in the gut and the environment. It should, however, be noted that the use of drugs generally has beneficial effects on the health and welfare of animals and on the efficiency of production. High health and welfare status of animals in more developed parts of the world is considered highly desirable by the public and tends to yield relatively inexpensive food products of high quality.

15 Acknowledgements

SAC and RRI are funded by the Scottish Office Agriculture Environment and Fisheries Dept.

Detection, Analysis and Risk Assessment of Cyanobacterial Toxins

S. G. BELL AND G. A. CODD

1 Introduction

Cyanobacteria (blue-green algae) are Gram-negative photosynthetic prokaryotes. In waters where nutrient enrichment has led to eutrophication,[1,2] toxic cyanobacterial species may dominate phytoplankton blooms.[3] During calm conditions, species of cyanobacteria which contain gas vesicles, being buoyant, rise to the surface of the water, and these can be manoeuvred to the margins of the waterbody by gentle breezes, where scums up to several centimetres thick may form. The occurrence of toxic cyanobacterial blooms and scums is a world-wide phenomenon, with examples in over 30 countries having been investigated in fresh, brackish or marine waters throughout Europe, North and South America, Australia, Africa, and Asia.[4-6]

A historical record exists of animal illnesses and fatalities as a result of cyanobacterial intoxication, dating back to at least 1853, including the first scientific documentation of the deaths of cattle, sheep, horses, pigs, and dogs after drinking from Lake Alexandrina in South Australia.[7,8] The lake contained the cyanobacterium *Nodularia spumigena*, a species capable of producing hepatotoxins. Since then, many incidents of animal intoxications attributable to toxic cyanobacteria have occurred, including cattle, sheep, pigs, horses, dogs, cats, monkeys, muskrats, squirrels, rhinoceros, fish, birds and invertebrates, and these have been reviewed.[6,9,10] Much of the evidence for the involvement of

1   C. S. Reynolds, *The Ecology of Freshwater Phytoplankton*, Cambridge University Press, Cambridge, 1990.
2   G. A. Codd, in *Proceedings of the Fourth Disaster Prevention and Limitation Conference*, ed. A. Z. Keller and H. C. Wilson, University of Bradford, 1992, Vol. 4, p. 33.
3   G. A. Codd and S. G. Bell, *J. Water Pollut. Control*, 1985, **84**, 225.
4   G. A. Codd, W. P. Brooks, L. A. Lawton and K. A. Beattie, in *Watershed '89. The Future for Water Quality in Europe*, ed. D. Wheeler, M. J. Richardson and J. Bridges, Pergamon Press, Oxford, 1989, vol. 2, p. 211.
5   W. E. Scott, *Water Sci. Technol.*, 1991, **23**, 175.
6   W. W. Carmichael and I. R. Falconer, in *Algal Toxins in Seafood and Drinking Water*, ed. I. R. Falconer, Academic Press, London, 1993, p. 187.
7   G. Francis, *Nature (London)*, 1878, **18**, 11.
8   G. A. Codd, D. A. Steffensen, M. D. Burch and P. D. Baker, *Aust. J. Mar. Freshwater Res.*, 1994, **45**, 731.
9   M. Schwimmer and D. Schwimmer, in *Algae, Man and the Environment*, ed. D. F. Jackson, Syracuse University Press, New York, 1968, p. 279.

cyanobacterial toxins in the earliest reported incidents of animal intoxications is circumstantial. Often the presence of a cyanobacterial bloom or scum in the water at the time of the incident was the only evidence, and on many occasions the toxicity of the cyanobacterial cells was not determined. However, more recent evidence has strongly implicated cyanobacterial toxins as the causative agents in animal and fish fatalities. A kill of approximately 2000 brown trout in Loch Leven and the River Leven, Scotland, in 1992 occurred shortly after the mass accumulation of the cyanobacterium *Anabaena flos-aquae*, which began to break open, releasing toxins into the water.[11] Laboratory work demonstrated that the cyanobacterial cells were lethal to mice, and that hepatotoxic microcystins were present in both cells and water of the loch and the river. Liver damage consistent with hepatotoxic intoxication was observed in dead and dying fish. Dog fatalities occurred at Loch Insh, Scotland, in the early 1990s after eating accumulations of benthic cyanobacteria of the *Oscillatoria/Phormidium* group.[12] Cyanobacterial cells were observed in the contents of the animals' stomachs, and the neurotoxin anatoxin-*a* was detected in the cyanobacterial cells and stomach contents.

Human illnesses and health problems have also been attributed to cyanobacterial toxins.[4,6,13] Incidents of human illness attributable to cyanobacterial toxins include: gastroenteritis disorders, the first case being published in 1931 after stagnation of the Ohio River in the USA resulted in a cyanobacterial bloom;[14] atypical pneumonia, resulting in the hospitalization of soldiers after canoeing in Rudyard Lake, England, which contained a toxic bloom of *Microcystis aeruginosa* at the time;[15] allergic and irritation reactions in freshwaters[16] and in marine water in the presence of *Lyngbya majuscula*;[17] liver diseases such as the outbreak of hepatoenteritis at Palm Island, Australia, when 139 children and 10 adults were affected after drinking from a supply reservoir containing the tropical cyanobacterium *Cylindrospermopsis raciborskii*;[18] and chronic liver damage amongst the population of Armidale, Australia, which had received its drinking water from a source contaminated with hepatotoxic *M. aeruginosa*.[19] An additional threat to human health posed by cyanobacterial hepatotoxins is that of liver tumour promotion. Recent studies in the People's Republic of China have demonstrated a higher incidence of primary liver cancer in a population taking drinking water from open ditches and ponds containing cyanobacterial hepatotoxins, compared with a neighbouring population receiving its drinking water from wells.[20]

[10] H. Utkilen, in *Photosynthetic Prokaryotes*, ed. N. H. Mann and N. G. Carr, Plenum Press, New York, 1992, p. 211.

[11] H. D. Rodger, T. Turnbull, C. Edwards and G. A. Codd, *J. Fish Dis.*, 1994, **17**, 177.

[12] C. Edwards, K. A. Beattie, C. M. Scrimgeour and G. A. Codd, *Toxicon*, 1992, **30**, 1165.

[13] S. G. Bell and G. A. Codd, *Rev. Med. Microbiol.*, 1994, **5**, 256.

[14] M. V. Veldee, *Am. J. Publ. Health*, 1931, **21**, 1227.

[15] P. C. Turner, A. J. Gammie, K. Hollinrake and G. A. Codd, *Br. Med. J.*, 1990, **300**, 1440.

[16] F. S. Soong, E. Maynard, K. Kirke and C. Luke, *Med. J. Aust.*, 1992, **156**, 67.

[17] R. E. Moore, I. Ohtani, B. S. Moore, C. B. de Koning, W.-Y. Yoshida, M. T. C. Runnegar and W. W. Carmichael, *Gazz. Chim. Ital.*, 1993, **123**, 329.

[18] A. T. C. Bourke, R. B. Hawes, A. Nielson and N. D. Stallman, *Toxicon*, 1983, **Suppl. 3**, 45.

[19] I. R. Falconer, A. M. Beresford and M. T. C. Runnegar, *Med. J. Aust.*, 1983, **1**, 511.

[20] S.-Z. Yu, in *Toxic Cyanobacteria: Current Status of Research and Management*, proceedings of an international workshop, ed. D. A. Steffensen and B. C. Nicholson, Australian Centre for Water Quality Research, Adelaide, 1994, p. 75.

The detection and analysis, including quantification, of cyanobacterial toxins are essential for monitoring their occurrence in natural and controlled waters used for agricultural purposes, potable supplies, recreation and aquaculture. Risk assessment of the cyanobacterial toxins for the protection of human and animal health, and fundamental research, are also dependent on efficient methods of detection and analysis. In this article we discuss the methods developed and used to detect and analyse cyanobacterial toxins in bloom and scum material, water and animal/clinical specimens, and the progress being made in the risk assessment of the toxins.

2   Production and Properties of Cyanobacterial Toxins

The genera of cyanobacteria which most frequently form blooms in European, including British, fresh and brackish waters all include species or strains capable of toxin production.[5,6,10,21] The toxins of cyanobacteria can be grouped into three broad categories: hepatotoxins, neurotoxins and lipopolysaccharide endotoxins. Three classes of cyanobacterial hepatotoxins are currently recognized: microcystins (cyclic heptapeptides), nodularins (cyclic pentapeptides) and cylindrospermopsin (a tricyclic alkaloid). Microcystins and nodularins are closely related and contain the novel amino acid 3-amino-9-methoxy-2,6,8-trimethyl-10-phenyldeca-4,6-dienoic acid (Adda) which, along with a cyclic structure of the toxins, is essential for toxicity. At least 50 variants of microcystin are now known, all conforming to the heptapeptide cyclic structure, variations occurring by substitutions in amino acid sites, by methylation or demethylation of the molecule, or by variations in the structure of the Adda moiety.[6,22] At present, five variants of nodularin have been identified.[22] Hepatotoxicoses resulting from exposure to microcystins or nodularins are the most common animal poisoning incidents reported in the literature, symptoms described including weakness, pallor, cold extremities, heavy breathing, diarrhoea, and vomiting. Death, as a result of liver haemorrhage, and a massive pooling of blood in the liver, usually occurs 2–24 hours after ingestion of the toxin or toxic cells. Microcystins are produced by axenic cultures of *Microcystis, Anabaena, Oscillatoria* and *Nostoc* strains, whereas nodularin has only been detected in laboratory cultures of *Nodularia* to date. Cylindrospermopsin is produced by strains of *Cylindrospermopsis raciborskii*[23] and *Umezakia natans*,[24] and is thought to be the toxin responsible for the hepatoenteritis outbreak on Palm Island, Australia. The survival time of laboratory mice dosed with cylindrospermopsin is much longer than that of animals dosed with microcystins or nodularins and, in addition to liver damage, damage to lungs, kidneys, adrenals and intestine also occurs.

The cyanobacterial neurotoxins, anatoxins and saxitoxins have been responsible

[21] G. A. Codd, C. Edwards, K. A. Beattie, L. A. Lawton, D. L. Campbell and S. G. Bell, in *Algae, Environment and Human Affairs*, ed. W. Wiessner, E. Schnepf and R. C. Starr, Biopress, Bristol, 1994, p. 1.

[22] K. L. Rinehart, M. Namikoshi and B. W. Choi, *J. Appl. Phycol.*, 1994, **6**, 159.

[23] P. R. Hawkins, M. T. C. Runnegar, A. R. B. Jackson and I. R. Falconer, *Appl. Environ. Microbiol.*, 1985, **50**, 1292.

[24] K.-I. Harada, I. Ohtani, K. Iwamoto, M. Suzuki, M. F. Watanabe, M. Watanabe and K. Terao, *Toxicon*, 1994, **32**, 339.

for animal, fish and bird poisoning incidents, but as yet have not been identified as sources of human illness, although saxitoxin and analogues produced by dinoflagellates is well-established as the cause of paralytic shellfish poisoning in humans. Anatoxin-*a* is a secondary amine alkaloid, produced by species of *Anabaena* and *Cylindrospermum*, as well as *Aphanizomenon flos-aquae* and benthic members of the *Oscillatoria/Phormidium* group.[12,25,26] Anatoxin-*a*, a neuromuscular blocking agent, is a postsynaptic, cholinergic, nicotinic agonist, which causes staggering, gasping, convulsions and muscle fasciculations in affected animals, in addition to causing arching of the neck backwards over the body (opisthotonos) in birds.[27] Death, as a result of respiratory arrest, occurs within a few minutes to a few hours. Anatoxin-*a*(*s*) is a *N*-hydroxyguanidine methyl phosphate ester, which causes hypersalivation in affected animals. It is a potent inhibitor of acetylcholinesterase and is approximately 10 times more potent than anatoxin-*a*.[27] Saxitoxin, neosaxitoxin and related gonyautoxins are produced by the cyanobacterial species *Aphanizomenon flos-aquae*[28] and *Anabaena circinalis*,[29] and are potent neurotoxins, blocking sodium channels and inhibiting nerve conduction. Animal fatalities due to ingestion of the saxitoxins, such as the large number of livestock along the Darling River, Australia, in 1991,[30] result from respiratory failure.

Lipopolysaccharide (LPS) endotoxins are characteristic Gram-negative outer-cell components which are produced by many cyanobacteria. Although LPS have been characterized and found to be toxic to laboratory animals after isolation from cyanobacteria, their toxicity to rodents is less potent than the endotoxins of enteric pathogens such as *Salmonella*.[31] Typical symptoms of animals suffering from LPS intoxication include vomiting, diarrhoea, weakness and death after hours rather than minutes.

In the summer of 1989, Rutland Water, the largest man-made lake in Western Europe and which supplies potable water to approximately 500 000 people in the East of England, contained a heavy bloom of *Microcystis aeruginosa*. By the end of the summer, a number of sheep and dogs had died after drinking from the bloom and concentrated scum. Analysis revealed that the cyanobacterial bloom material was toxic to laboratory mice,[32] and that rumen contents from a poisoned sheep contained five microcystin variants.[33] Microcystins were detected in waters used for recreation in Australia at concentrations greater than 1 mg per

[25] W. W. Carmichael, D. F. Biggs and P. R. Gorham, *Science (Washington, DC)*, 1975, **187**, 542.

[26] K. Sivonen, K. Himberg, R. Luukkainen, S. Niemela, G. Poon and G. A. Codd, *Toxic. Assess.*, 1989, **4**, 339.

[27] W. W. Carmichael, *J. Appl. Bacteriol.*, 1992, **72**, 445.

[28] N. A. Mahmood and W. W. Carmichael, *Toxicon*, 1986, **24**, 175.

[29] A. R. Humpage, J. Rositano, A. H. Bretag, R. Brown, P. D. Baker, B. C. Nicholson and D. A. Steffensen, *Aust. J. Mar. Freshwater Res.*, 1994, **45**, 761.

[30] L. Bowling, *The Cyanobacterial (Blue-Green Algal) Bloom in the Darling/Barwon River System*, Report No. 92.074, NSW Department of Water Resources, Technical Services Division, Parramatta, 1991.

[31] S. Raziuddin, H. W. Siegelman and T. G. Tornabene, *Eur. J. Biochem.*, 1983, **137**, 333.

[32] National Rivers Authority, *Toxic Blue-Green Algae: The Report of the National Rivers Authority*, Water Quality Series No. 2, National Rivers Authority, London, 1990.

[33] L. A. Lawton, C. Edwards, K. A. Beattie, S. Pleasance, G. J. Dear and G. A. Codd, *Natural Toxins*, 1995, **3**, 50.

litre, after lysis of the cells of the *Microcystis aeruginosa* present in the water by algicide treatment in 1992.[34] Thus the potential for cyanobacterial toxins to cause human ill-health exists in potable waters and recreational waters.

3 Detection and Analysis of Cyanobacterial Toxins

The detection and quantification of cyanobacterial toxins quoted in the above examples required methods which have been undergoing rapid development in recent years, and as the need for greater understanding of the properties and occurrence of the toxins continues to grow, these are continuing to be developed. This has resulted in methods of cyanobacterial toxin detection which are more sensitive, quantitative, reliable, specific and humane. Many of these methods are presented and discussed in the proceedings of a recent conference.[35]

In order to achieve maximal detection efficiency, a good understanding of the intra- and extracellular localization, stability and fate of the toxins is required. These requirements include knowledge of the temperature-stability and binding characteristics of the toxins during sampling and laboratory processing procedures. Microcystins were only detected at low concentrations (approximately $5 \mu g \, l^{-1}$) in water during a bloom of *M. aeruginosa* in Lake Centenary, Australia, prior to algicide treatment, but after treatment, high concentrations of the toxin ($mg \, l^{-1}$) were detected in the water.[34] In contrast, Rapala *et al.*[36] detected almost 20% of the total anatoxin-*a* in the growth medium of a healthy growing culture of *Aphanizomenon flos-aquae*. Thus, in a naturally occurring bloom of cyanobacteria, methods for detection of toxins should make provision for analysis of both intra- and extracellular components. Extracellular microcystin was shown to be degraded in natural waters after a lag-period of approximately 9 days[34] and, in the laboratory, degradation of microcystins-LR and -YR was observed after a lag-period of 10–20 days in decomposing cyanobacterial cultures.[37] Microcystin-LR is a highly stable molecule, in that it is unaffected by boiling for up to 1 hour or pH 1–10 for long periods, and does not break down in sterile solution at room temperature for over 1 year.[38] Furthermore, microcystin-LR does not appear to attach to mineral particulates or phytoplankton in a water column, although it does bind to some plastics but not to glass on serial transfer. Thus, when devizing sampling and handling procedures for the quantitative analysis of microcystins it would appear pertinent to use glass containers where possible, and storage should be under conditions where bacterial activity is minimized. These factors should be taken into consideration when handling cyanobacterial samples for determination of other toxins.

Methods for the detection and analysis of cyanobacterial toxins fall into two

[34] G. J. Jones and P. T. Orr, *Water Res.*, 1994, **28**, 871.

[35] G. A. Codd, T. M. Jefferies, C. W. Keevil and E. Potter, *Detection Methods for Cyanobacterial Toxins*, Royal Society of Chemistry, Cambridge, 1994.

[36] J. Rapala, K. Sivonen, R. Luukkainen, and S. Niemela, *J. Appl. Phycol.*, 1993, **5**, 581.

[37] M. F. Watanabe, K. Tsuji, Y. Watanabe, K.-I. Harada and M. Suzuki, *Natural Toxins*, 1992, **1**, 48.

[38] G. A. Codd and S. G. Bell, *Occurrence, Fate and Behaviour of Cyanobacterial (Blue-Green Algal) Hepatotoxins*, R & D Project Record 271/7/A, National Rivers Authority, Bristol, 1995.

broad categories: biological systems, including bioassays and biochemical methods, and physicochemical methods.

Biological Assays

The mouse bioassay is the longest-established, and was in the past the most extensively used, method for determination of cyanobacterial toxins, especially as an initial screen for detecting the presence of toxins in cyanobacterial blooms or cultures. Its implementation relies on the use of an in-bred strain of laboratory animals which are dosed orally or intraperitoneally. Quantification of toxicity is still determined as the concentration of test material which is lethal to 50% of the population of animals dosed (LD_{50}) or, more appropriately today, as the minimum lethal dose for the population (MLD_{100}). The mouse bioassay is a rapid, low-technology test, which requires licensed, specially trained operators. It is useful as an initial screen, in that symptoms and survival times of dosed animals can be used as indicators of the type of cyanobacterial toxin involved. However, cloning and housing animals under strictly regulated and licensed conditions is expensive, and there are objections to the use of this bioassay for humane and moral reasons.

Other bioassay methods have been employed and evaluated as alternatives or additions to the mouse bioassay, including *Daphnia, Artemia*, bacterial and cytotoxicity bioassays. In 1981, Lampert demonstrated that *M. aeruginosa* was toxic to *Daphnia pulicaria*, in that survival of new-born daphnids and filtering rates were reduced,[39] and recently it has been shown that *Daphnia hyalina* and *Daphnia pulex* are also susceptible when exposed to toxic *Microcystis* and purified microcystins.[40,41] Typically, *Daphnia* are exposed to toxicants for 24–48 hours and toxicity is expressed as the LC_{50} value (concentration which is lethal to 50% of the population). However, different species of *Daphnia* have differing susceptibilities to microcystins due to differences in feeding responses to toxic cells and to differing sensitivities to the purified toxins.[40,41] This suggests that daphnids may evolve physiological and behavioural adaptations to coexist with toxic cyanobacteria. To overcome problems of variation in assay results, the cultures of *Daphnia* used should be standardized, as suggested by Baird *et al.*[42] who recommended the use of a single species, and strictly controlled, consistent culturing conditions including: controlled stocking densities, use of synthetic media, use of axenic cultures of algae as the food source, and consistency in the quantity of food given to the daphnids.

The brine shrimp (*Artemia salina*) has been evaluated as an alternative to the mouse bioassay for use in cyanobacterial toxicity screening assays.[43,44] As in the

[39] W. Lampert, *Verh. Int. Verein. Limnol.*, 1981, **21**, 1436.

[40] L. A. Lawton, S. P. Hawser, K. Jamel Al-Layl, K. A. Beattie, C. MacKintosh and G. A. Codd, in *Proceedings of the Second Biennial Water Quality Symposium*, ed. G. Castillo, V. Campos and L. Herrera, University of Santiago, Chile, 1990, p. 83.

[41] W. R. DeMott, Q.-X. Zhang and W. W. Carmichael, *Limnol. Oceanogr.*, 1991, **36**, 1346.

[42] D. J. Baird, A. M. V. M. Soares, A. Girling, I. Barber, M. C. Bradley and P. Callow, in *Proceedings of the First European Conference on Ecotoxicology, Copenhagen*, ed. H. Lokke, H. Tyle and F. BroRasmussen, Lyngby, 1988, p. 144.

[43] J. Kiviranta, K. Sivonen and S. I. Niemela, *Environ. Toxicol. Water Qual.*, 1991, **6**, 423.

Daphnia assay, the brine shrimps are exposed to different concentrations of toxicant, and the toxicity is expressed as the LC_{50} value. Extracts of cyanobacterial blooms and laboratory cultures, containing microcystins or anatoxin-*a*, have been found to be toxic towards brine shrimp,[43] and fractionation of such extracts resulted in brine shrimp fatalities only with fractions containing microcystins.[44]

A number of assays involving the use of bacteria for cyanobacterial toxin detection and quantification have been investigated. These include assays for the inhibition of bioluminescence, pigment production and growth. Microtox is a commercially available assay system which utilizes *in vivo* the bioluminescence of *Photobacterium phosphoreum*. The toxicant is incubated with the bacteria and the semi-automated system detects the luminescence and calculates the EC_{50} value (effective concentration causing a 50% reduction in bioluminescence). Crude extracts of hepatotoxic blooms and cultures of cyanobacteria and purified microcystin-LR inhibit Microtox luminescence.[45] However, the greatest inhibition was later shown to be associated with some factor(s) other than the microcystins in microcystin-containing and non-microcystin-containing cyanobacterial fractions.[44] Neurotoxic *Anabaena* samples were also tested in the Microtox system and, as with microcystins, the inhibition of luminescence did not correlate with the abundance of anatoxin-*a*.[46] Production of the pigment prodigiosin by *Serratia marcescens* was inhibited by extracts of toxic strains of *M. aeruginosa* and *Aph. flos-aquae*.[47] Inhibition by the cyanobacterial extracts was dose-dependent, but no correlation was observed with purified microcystin, indicating that the inhibition was not primarily caused by the peptide toxin. Recently, effects of microcystins and anatoxin-*a* on the growth of *Pseudomonas putida* was investigated by Lahti *et al.*[46] Turbidity of cultures of the bacteria was measured in the presence and absence of the toxin, over a 16-hour period, and inhibition compared with a control (no toxin present) was monitored. Microcystin fractions enhanced the turbidity of the bacterial cultures, whereas anatoxin-*a* fractions inhibited bacterial growth, but this did not correlate with increasing toxin concentration.

In vitro cytotoxicity assays using isolated cells have been applied intermittently to cyanobacterial toxicity testing over several years.[48–51] Cells investigated for suitability in cyanobacterial toxin assays include primary liver cells (hepatocytes) isolated from rodents and fish, established permanent mammalian cell lines, including hepatocytes, fibroblasts and cancerous cells, and erythrocytes. Earlier work suggested that extracts from toxic cyanobacteria disrupted cells of established lines and erythrocytes,[48] but studies with purified microcystins revealed no alterations in structure or ion transport in fibroblasts or erythrocytes,

[44] D. L. Campbell, L. A. Lawton, K. A. Beattie and G. A. Codd, *Environ. Toxicol. Water Qual.*, 1994, **9**, 71.

[45] L. A. Lawton, D. L. Campbell, K. A. Beattie and G. A. Codd, *Lett. Appl. Microbiol.*, 1990, **11**, 205.

[46] K. Lahti, J. Ahtiainen, J. Rapala, K. Sivonen and S. I. Niemela, *Lett. Appl. Microbiol.*, 1995, **21**, 109.

[47] R. Dierstein, I. Kaiser and J. Weckesser, *Syst. Appl. Microbiol.*, 1989, **12**, 244.

[48] W. O. K. Grabow, W. C. DuRandt, O. W. Prozesky and W. E. Scott, *Appl. Environ. Bacteriol.*, 1982, **43**, 225.

[49] G. A. Codd, W. P. Brooks, I. M. Priestley, G. K. Poon, S. G. Bell and J. K. Fawell, *Toxic. Assess.*, 1989, **4**, 499.

[50] J. E. Eriksson, H. Hagerstrand and B. Isomaa, *Biochim. Biophys. Acta*, 1987, **930**, 304.

[51] K. Henning, H. Meyer, G. Kraatz-Wadsack and J. Cremer, *Curr. Microbiol.*, 1992, **25**, 129.

suggesting that other compounds in the crude extracts may have resulted in earlier inhibitory effects.[50] Recently, a substance other than microcystin has been isolated from *M. aeruginosa* that caused lysis of established cell lines, including liver cells.[51] However, purified microcystins have been found to cause structural alteration and lysis in isolated hepatocytes from rats[50] and rainbow trout,[52] which is indicative of an active transport system for microcystins possessed by hepatocytes.[53] An assay for sodium channel-blocking toxins (PSP toxins) such as saxitoxin is available using mouse neuroblastoma cells.[54] These cells die in the presence of ouabain and veratridine which open sodium channels, but when sodium channel-blocking toxins are present the lethal effect is nullified. The assays are carried out in microtitre plates over 2 days, and the vital stain, neutral red, added to detect viable cells, the colour being measured in a plate reader. Although the neuroblastoma assay responds to sodium channel-blocking toxins, it is not inferred to be specific for saxitoxin and the related cyanobacterial and algal neurotoxins.

Antibodies have been raised against cyanobacterial toxins and subsequently used in immunoassays for their detection. The pioneering work in this field was carried out by Kfir *et al.*,[55] who raised monoclonal antibodies against microcystin-LA. These antibodies were found to be specific for the toxin variant. Brooks and Codd[56] raised polyclonal antibodies against microcystins isolated from *M. aeruginosa*, which were used in an enzyme-linked immunosorbent assay (ELISA). Positive results were only found with extracts of toxic *M. aeruginosa* laboratory cultures, but not with extracts of other species of cyanobacteria which also contained microcystins. Chu *et al.*[57] also raised polyclonal antibodies against a microcystin variant, microcystin-LR, which when used in ELISA assays were found to cross-react with microcystin-LR and other microcystin variants.[58] Recent work by Nagata *et al.* has also produced monoclonal antibodies against microcystin-LR, and these were shown to cross-react with a number of microcystin variants and nodularin in an ELISA format.[59] Polyclonal antibodies have been raised against saxitoxin and neosaxitoxin, and these have been applied to ELISA detection systems.[60,61] In both of these projects the polyclonal antibodies cross-reacted with saxitoxin and neosaxitoxin.

At the molecular level, microcystins are potent inhibitors of protein phosphatases 1 and 2A.[62] The activity of protein phosphatases can be determined by measuring

[52] J. E. Eriksson, D. M. Toivola, M. Reinikainen, C. M. I. Rabergh and J. A. O. Meriluoto, in *Detection Methods for Cyanobacterial Toxins*, ed. G. A. Codd, T. M. Jefferies, C. W. Keevil and E. Potter, Royal Society of Chemistry, Cambridge, 1994, p. 75.

[53] M. T. C. Runnegar, R. G. Gerdes and I. R. Falconer, *Toxicon*, 1991, **29**, 43.

[54] S. Gallacher and T. H. Birkbeck, *FEMS Microbiol. Lett.*, 1992, **92**, 101.

[55] R. Kfir, E. Johannsen and D. P. Botes, in *Mycotoxins and Phycotoxins. Bioactive Molecules*, ed. P. S. Steyn and R. Vleggaar, Elsevier Science, Amsterdam, 1986, vol. 1, p. 377.

[56] W. P. Brooks and G. A. Codd, *Environ. Technol. Lett.*, 1988, **9**, 1343.

[57] F. S. Chu, X. Huang, R. D. Wei and W. W. Carmichael, *Appl. Environ. Microbiol.*, 1989, **55**, 1928.

[58] F. S. Chu, X. Huang and R. D. Wei, *J. Assoc. Off. Anal. Chem.*, 1990, **73**, 451.

[59] S. Nagata, H. Soutome, T. Tsutsumi, A. Hasegawa, M. Sekijima, M. Sugamata, K.-I. Harada, M. Suganuma and Y. Ueno, *Natural Toxins*, 1995, **3**, 78.

[60] F. S. Chu, X. Huang and S. Hall, *J. AOAC Int.*, 1992, **75**, 341.

[61] E. Usleber, E. Scheider, G. Terplan and M. V. Laycock, *Food Add. Contam.*, 1995, **12**, 405.

[62] C. MacKintosh, K. A. Beattie, S. Klumpp, P. Cohen and G. A. Codd, *FEBS Lett.*, 1990, **264**, 187.

the release of phosphate from phosphorylated protein substrates, and microcystins can be detected by measuring the inhibition of phosphate release.[63] The microcystins detected can be quantified by comparing the level of inhibition of protein phosphatase to a standard curve constructed with microcystin standards of known concentration. In order to measure the amount of phosphate released, assay systems have used ^{32}P-labelled substrate,[63–65,33]P-labelled substrate,[66] and a colorimetric assay involving the chromogenic substrate *p*-nitrophenol phosphate.[67] When analysis of cyanobacterial extracts for microcystins is performed using protein phosphatase inhibition assays, the influence of endogenous phosphatase or proteolytic activity of the cyanobacteria should be considered. Endogenous enzyme activity would result in cleavage of phosphate from a phosphorylated protein substrate, resulting in underestimated detection of microcystins. To overcome this problem, it has been suggested that the endogenous phosphatase activity of the cyanobacteria should be quantified, or a phosphorylated peptide substrate instead of a protein substrate should be used.[68] A further drawback of the protein phosphatase inhibition assay is that it is not specific for microcystins, in that other non-cyanobacterial toxins also inhibit protein phosphatases.[63] However, if a test sample does inhibit protein phosphatase activity it should be considered toxic, and thus the assay is useful as a toxicity screen.

An enzymatic assay can also be used for detecting anatoxin-*a*(*s*).[69] This toxin inhibits acetylcholinesterase, which can be measured by a colorimetric reaction, *i.e.* reaction of the acetyl group, liberated enzymatically from acetylcholine, with dithiobisnitrobenzoic acid. The assay is performed in microtitre plates, and the presence of toxin detected by a reduction in absorbance at 410 nm when read in a plate reader in kinetic mode over a 5 minute period. The assay is not specific for anatoxin-*a*(*s*) since it responds to other acetylcholinesterase inhibitors, *e.g.* organophosphorus pesticides, and would need to be followed by confirmatory tests for the cyanobacterial toxin.

Physicochemical Assays

In addition to detection of toxicity in samples containing cyanobacteria and/or their toxins (*i.e.* screening), quantification and identification of the toxins present are necessary on occasions. Physicochemical methods of toxin analysis fulfil both these roles, often requiring a comparison of the test sample with purified

[63] C. MacKintosh and R. W. MacKintosh, in *Detection Methods for Cyanobacterial Toxins*, ed. G. A. Codd, T. M. Jefferies, C. W. Keevil and E. Potter, Royal Society of Chemistry, Cambridge, 1994, p. 90.

[64] J. Chaivimol, U. K. Swoboda and C. S. Dow, in *Detection Methods for Cyanobacterial Toxins*, ed. G. A. Codd, T. M. Jefferies, C. W. Keevil and E. Potter, Royal Society of Chemistry, Cambridge, 1994, p. 172.

[65] C. Edwards, L. A. Lawton and G. A. Codd, in *Detection Methods for Cyanobacterial Toxins*, ed. G. A. Codd, T. M. Jefferies, C. W. Keevil and E. Potter, Royal Society of Chemistry, Cambridge, 1994, p. 175.

[66] A. Sahin, F. G. Tencalla, D. R. Dietrich, K. Mez and H. Naegeli, *Am. J. Vet. Res.*, 1995, **56**, 1110.

[67] J. An and W. W. Carmichael, *Toxicon*, 1994, **12**, 1495.

[68] A. T. R. Sim and L.-M. Mudge, in *Detection Methods for Cyanobacterial Toxins*, ed. G. A. Codd, T. M. Jefferies, C. W. Keevil and E. Potter, Royal Society of Chemistry, Cambridge, 1994, p. 100.

[69] G. A. Codd, *The Toxicity of Benthic Blue-Green Algae in Scottish Freshwaters*, Report No. FR/SC 0012, Foundation for Water Research, Marlow, 1995.

standards. Such methods include high-performance liquid chromatography (HPLC), mass spectrometry (MS) and electrophoretic methods.

HPLC analysis of microcystins contained within cyanobacteria requires extraction of the toxin, and has been carried out since the early 1980s. Gregson and Lohr[70] used ultrasonicated lyophilized cyanobacterial material which had been resuspended in carbonate buffer, followed by ion exchange and gel permeation chromatography, and finally HPLC using a reversed-phase C_{18} column. Since then, extraction procedures have included 5% butanol/20% methanol (v/v) followed by solid-phase extraction through C_{18} cartridges,[71] extraction with 5% acetic acid followed by octadecylinized (ODS) silica gel clean-up,[72] and extraction with methanol.[73] The choice of extraction agents is an important factor when analyzing cyanobacterial samples for microcystin content, depending on the hydrophobicity of the microcystins present. Lawton *et al.*[73] demonstrated that the recovery of the most hydrophobic microcystins present in their samples, *e.g.* microcystins-LW and -LF, was poor when extracted with 5% acetic acid compared with recoveries obtained when using methanol. Although reversed-phase C_{18} columns are still favoured for analysis of microcystins, mobile phases used vary and include methanol–trifluoroacetic acid (TFA)/water,[72] and acetonitrile/TFA–water/TFA.[73] UV detection of microcystins is carried out at 238 nm, and retention times and areas of peaks obtained can be compared with purified standards to estimate identity and quantity, respectively.

A powerful tool now employed is that of diode array detection (DAD).[73] This function allows peaks detected by UV to be scanned, and provides a spectral profile for each suspected microcystin. Microcystins have characteristic absorption profiles in the wavelength range 200–300 nm, and these can be used as an indication of identity without the concomitant use of purified microcystin standards for all variants.[73] A HPLC–DAD analytical method has also been devised for measurement of intracellular and extracellular microcystins in water samples containing cyanobacteria.[73] This method involves filtration of the cyanobacteria from the water sample. The cyanobacterial cells present on the filter are extracted with methanol and analysed by HPLC. The filtered water is subjected to solid-phase clean-up using C_{18} cartridges, before elution with methanol and then HPLC analysis.

HPLC analysis of anatoxin-*a* was first carried out by Astrachan and Archer,[74] who extracted the toxin from *Anabaena flos-aquae* using chloroform followed by hydrochloric acid. The HPLC analysis was carried out on an ODS column using hypochlorate–methanol. Other systems used since include acetic acid extraction and analysis on a reversed-phase C_{18} column using methanol–water mobile phase,[75] and extraction in water after ultrasonication and analysis on reversed-phase

[70] R. P. Gregson and R. R. Lohr, *Comp. Biochem. Physiol., C: Comp. Pharmacol. Toxicol.*, 1983, **74**, 413.

[71] W. P. Brooks and G. A. Codd, *Lett. Appl. Microbiol.*, 1986, **2**, 1.

[72] K.-I. Harada, M. Suzuki, A. M. Dahlem, V. R. Beasley, W. W. Carmichael and K. L. Rinehart, *Toxicon*, 1988, **26**, 433.

[73] L. A. Lawton, C. Edwards and G. A. Codd, *Analyst*, 1994, **119**, 1525.

[74] N. B. Astrachan and B. G. Archer, in *The Water Environment: Algal Toxins and Health*, ed. W. W. Carmichael, Plenum Press, New York, 1981, p. 437.

[75] K.-I. Harada, Y. Kimura, K. Ogawa, M. Suzuki, A. M. Dahlem, V. R. Beasley and W. W. Carmichael, *Toxicon*, 1989, **27**, 1289.

C_{18} with an acetonitrile/phosphate buffer mobile phase.[12] PSP toxins have been extracted from *Anabaena circinalis*[29] and from freshwater mussels fed with a culture of the same cyanobacterium.[76] Acid extraction of cyanobacterial cells and mussels was carried out, followed by analysis on C_8 reversed-phase columns, and derivatization of the PSP toxins to fluorescent compounds, required for visualization, was carried out post-column. Thin-layer chromatography (TLC) and high-performance thin-layer chromatography (HPTLC), where a reversed-phase C_{18} material was coated onto glass plates, have been used to analyse anatoxin-*a* and microcystins.[77,78]

Mass spectrometry has been used to characterize microcystins using the method of fast-atom bombardment (FAB) ionization and MS/MS.[79,80] Anatoxin-*a* has been analysed by MS in combination with gas chromatography in bloom and water samples,[81] and in benthic cyanobacterial material and stomach contents of poisoned animals.[12] Recently, liquid chromatography (LC) linked to MS has been employed to analyse microcystins, where FAB-MS[82] and atmospheric-pressure ionization (API-MS)[83] have been used, and anatoxin-*a*, where thermospray (TSP-MS) was used.[84]

Early electrophoretic methods for the detection of cyanobacterial toxins include preparative paper electrophoresis of microcystins, which were concentrated by gel-permeation and ion exchange chromatography before resolution by electrophoresis.[85] Since then, capillary electrophoresis (CE), which separates compounds on small diameter silica columns according to differences in their molecular size and charge, has been used to analyse microcystins[86] and anatoxins.[87]

Sample Analysis

Analysis of environmental and laboratory samples for cyanobacterial toxins falls into a number of categories: screening for the presence of the toxins, identification of toxins present and quantification of toxicity and/or toxins. Screening should provide an indication of the presence, in all cases, of the toxin sought; it should

[76] A. P. Negri and G. J. Jones, *Toxicon*, 1995, **33**, 667.

[77] I. Ojanpera, E. Vuori, K. Himberg, M. Waris and K. Niinivaara, *Analyst*, 1991, **116**, 265.

[78] K. Jamel Al-Layl, G. K. Poon, and G. A. Codd, *J. Microbiol. Meth.*, 1988, **7**, 251.

[79] T. Krishnamurthy, L. Szafraniec, D. F. Hunt, J. Shabanowitz, I. Yates, C. R. Hauer, W. W. Carmichael, O. M. Skulberg, G. A. Codd and S. A. Missler, *Proc. Natl. Acad. Sci. USA*, 1989, **86**, 770.

[80] K.-I. Harada, T. Mayumi, T. Shimada, M. Suzuki, F. Kondo and M. F. Watanabe, *Tetrahedron Lett.*, 1993, **34**, 6091.

[81] K. Himberg, *J. Chromat.*, 1989, **481**, 358.

[82] F. Kondo, Y. Ikai, H. Oka, N. Ishikawa, M. F. Watanabe, M. Watanabe, K.-I. Harada and M. Suzuki, *Toxicon*, 1992, **30**, 227.

[83] C. Edwards, L. A. Lawton, K. A. Beattie, G. A. Codd, S. Pleasance and G. J. Dear, *Rapid Commun. Mass Spectrom.*, 1993, **7**, 714.

[84] K.-I. Harada, H. Nagai, Y. Kimura, M. Suzuki, H.-D. Park, M. F. Watanabe, R. Luukkainen, K. Sivonen and W. W. Carmichael, *Tetrahedron*, 1993, **49**, 9251.

[85] T. C. Elleman, I. R. Falconer, A. R. B. Jackson and M. T. C. Runnegar, *Aust. J. Biol. Sci.*, 1978, **31**, 209.

[86] M. P. Boland, M. A. Smillie, D. Z. X. Chen and C. F. B. Holmes, *Toxicon*, 1993, **31**, 1393.

[87] T. M. Jefferies, G. Brammer, A. Zoton, P. A. Brough and T. Gallagher, in *Detection Methods for Cyanobacterial Toxins*, ed. G. A. Codd, T. M. Jefferies, C. W. Keevil and E. Potter, Royal Society of Chemistry, Cambridge, 1994, p. 34.

not detect any false negative samples, but need not necessarily be specific for any individual or group of cyanobacterial toxins. Ideally, screening methods should be rapid and easy to use. Table 1 presents the characteristics of biological assays for cyanobacterial toxins and indicates their suitability for use as screening methods.

After screening for toxicity, identification and/or quantification assays may need to be carried out if the screening method is not specific for the cyanobacterial toxin(s) under investigation. Suitable assays for these purposes include the physicochemical assays, HPLC, MS, and CE, and to some extent the immunoassays and protein phosphatase inhibition assays summarized in Section 2.

The ability to identify and quantify cyanobacterial toxins in animal and human clinical material following (suspected) intoxications or illnesses associated with contact with toxic cyanobacteria is an increasing requirement. The recoveries of anatoxin-*a* from animal stomach material[12] and of microcystins from sheep rumen contents[33] are relatively straightforward. However, the recovery of microcystin from liver and tissue samples cannot be expected to be complete without the application of proteolytic digestion and extraction procedures. This is likely because microcystins bind covalently to a cysteine residue in protein phosphatase.[88,89] Unless an effective procedure is applied for the extraction of covalently bound microcystins (and nodularins), then a negative result in analysis cannot be taken to indicate the absence of toxins in clinical specimens. Furthermore, any positive result may be an underestimate of the true amount of microcystin in the material and would only represent free toxin, not bound to the protein phosphatases. Optimized procedures for the extraction of bound microcystins and nodularins from organ and tissue samples are needed.

4   Risk Assessment of Cyanobacterial Toxins

In order to counter the hazards presented to health by cyanobacterial toxins, management actions concerning potable and recreational waters are required. These actions include risk assessment and monitoring programmes[13] which rely on sensitive, accurate toxin analysis methods.

A requirement for potable waters affected by cyanobacterial toxins is risk assessment. To date, owing to the lack of data, such assessments have been performed only for the presence of microcystins in drinking water. In Australia, a provisional safety guideline for the maximum concentration of microcystin in drinking water of $1\ \mu g\,l^{-1}$ is presently recommended.[6] This value was determined using data from oral administration of microcystins to mice[90] and on subchronic oral toxicity to pigs.[91] Applying safety factors, the calculated toxin concentrations for a 60 kg person drinking 2 litres of water per day were $1.5\ \mu g\,l^{-1}$ and $0.84\ \mu g\,l^{-1}$, respectively. A further risk assessment of microcystins in drinking

88 J. Goldberg, H.-B. Huang, Y.-G. Kwon, P. Greengard, A.C. Nairn and J. Kuriyan, *Nature (London)*, 1995, **376**, 745.
89 R. W. MacKintosh, K. N. Dalby, D. G. Campbell, P. T. W. Cohen, P. Cohen and C. MacKintosh, *FEBS Lett.*, 1995, **371**, 236.
90 I. R. Falconer, J. V. Smith, A. R. B. Jackson, A. Jones and M. T. C. Runnegar, *J. Toxicol. Environ. Health*, 1988, **24**, 291.
91 I. R. Falconer, M. D. Burch, D. A. Steffensen, M. Choice and O. R. Coverdale, *J. Environ. Toxicol. Water Qual.*, 1994, **9**, 131.

Table 1 Characteristics of biological assays for cyanobacterial toxins

| Method | Duration | Logistics* | Toxin identification | | | Suitable for screening |
			Mode of action	Specificity	Sensitivity[†]	
Mouse	mins–hrs	lip, lr, sto	yes	no	yes	yes (rapid)
Daphnia	days	lip, easy	no	no	yes	yes (slow)
Artemia	days	easy	no	no	yes	yes (slow)
Microtox	mins	easy, ser	no	no	no	no
Cytotoxicity	mins–days	lip, sto,ser	yes[‡]	no	yes	yes[‡] (variable)
Immunoassays	mins–hrs	easy, ser[§]	yes[¶]	yes[¶]	yes	yes (rapid)
PPase inhibition						
1. radioisotopic	min–hrs	lip, lr, sto, ser	yes[‖]	no	yes	yes[‖] (rapid)
2. colorimetric	mins	easy, ser	yes[‖]	no	yes	yes[‖] (rapid)
Cholinesterase	mins	easy, ser	yes**	no	yes	yes** (rapid)

*lip = labour intensive preparation, lr = licence required, sto = specially trained operators, easy = easy to use, ser = special equipment required.
[†]Sensitive to toxins, in this case means that the assay presents no false negative results.
[‡]Primary hepatocytes can elucidate hepatotoxins, and mouse neuroblastoma cells can elucidate sodium channel-blocking neurotoxins; therefore these assays can be used to screen for the appropriate toxins.
[§]No special equipment is required for some assay formats.
[¶]Specificity of the assay depends on the specificity (cross-reactivity) of the antibodies.
[‖]Of the known cyanobacterial toxins, only hepatotoxins are detected and are, therefore, able to be screened for by protein phosphatase inhibition.
**Of the known cyanobacterial toxins only anatoxin-*a(s)* is detected and is, therefore, able to be screened for by acetylcholinesterase inhibition.

water, using data from a 90-day oral dose study of microcystins to mice, resulted in a guidance value of 0.5 µg microcystin per litre of water.[92] This calculation took into account the no observed adverse effect level (NOAEL) of microcystin to mice, and applied safety factors including evidence for tumour promotion, for a 70 kg person consuming 1.5 litres of water daily. These risk assessment values for microcystins should be regarded as provisional guidelines and should be kept under review, as more data on tumour promotion and human health effects, including epidemiological studies, become available.

It is obvious from the provisional risk assessment values for microcystins, and, being of the same order of magnitude of mammalian toxicity, similar values may be calculated for the cyanobacterial neurotoxins, that sensitive detection methods are required to detect these low concentrations of toxins. Of the biological methods of detection discussed earlier, the mouse and invertebrate bioassays are not sensitive enough without concentration of water samples, in that they are only able to detect mg of microcystins per litre. Only the immunoassays $(ng-\mu g\,l^{-1})$ and the protein phosphatase inhibition assays $(ng\,l^{-1})$

[92] T. Kuiper-Goodman, S. Gupta, H. Combley and B. H. Thomas, in *Toxic Cyanobacteria: Current Status of Research and Management*, proceedings of an international workshop, eds. D. A. Steffensen and B. C. Nicholson, Australian Centre for Water Quality Research, Adelaide, 1994, p. 67.

are sufficiently sensitive to detect low levels of microcystin in water without a concentration step.

Not all cyanobacterial blooms and scums contain detectable levels of toxins. Indeed, the incidence of toxicity detection by mouse bioassay, and toxin detection by HPLC among environmental samples, ranges from about 40% to 75%.[4,38,93] However, in view of this high occurrence, it is the policy of regulatory authorities and water supply operators in some countries to assume that blooms of cyanobacteria are toxic until tested and found to be otherwise. In the absence of available analytical facilities or expertise or for logistical reasons, this precautionary principle should be regarded as sensible and prudent.

In reality, the demands upon most freshwaters, whether from human potable supply, animal watering, aquaculture, recreation, or amenity are increasing, and a need exists to monitor the types, location and levels of cyanobacterial toxins.

In summary, the toxicity testing of cyanobacterial blooms and scums can be achieved by a number of methods. These differ according to ease of use, duration, sensitivity and specificity, and the choice of method depends on the application, be it a toxicity screen or a quantitative or confirmation of identity method. The requirement for widely available, low unit cost, sensitive screening methods is likely to increase as the incidence of eutrophication and toxic cyanobacterial bloom production increases in freshwaters,[1-6,94] and as the recognition of this incidence and of the health significance of the toxins develops among waterbody regulators, operators and users.

[93] P. D. Baker and A. R. Humpage, *Aust. J. Mar. Freshwater Res.*, 1994, **45**, 773.
[94] D. Stanners and P. Bourdeau, *Europe's Environment. The Dobris Assessment*, European Environment Agency, Copenhagen, 1995.

Subject Index

125